GEOHAZARDS CAUSED BY HUMAN ACTIVITY

Edited by **Arvin Farid**

Geohazards Caused by Human Activity

http://dx.doi.org/10.5772/61670
Edited by Arvin Farid

Contributors

Maria Rosaria Senatore, Maddalena Falco, Agostino Meo, Moung Jin Lee, Kuan-Tsung Chang, Edward H. Wang, Chih-Ping Kuo, Min-Cheng Teng, Jin-King Liu, Franco Oboni, Ahmed Youssef, Milorad Jovanovski, Igor Peshevski

Published by InTech

Janeza Trdine 9, 51000 Rijeka, Croatia

Publishing Process Manager Edi Lipovic

Technical Editor SPi Global

Cover InTech Design team

First published November, 2016

Printed in Croatia

Additional hard copies can be obtained from orders@intechopen.com

Geohazards Caused by Human Activity, Edited by Arvin Farid

p. cm.

Print ISBN 978-953-51-2801-4
Online ISBN 978-953-51-2802-1

For EU product safety concerns:
IN TECH d.o.o., Prolaz Marije Krucifikse Kozulić 3, 51000 Rijeka, Croatia,
info@intechopen.com or visit our website at intechopen.com.

Meet the editor

Dr. Arvin Farid is an associate professor and the graduate coordinator of the Civil Engineering Department at Boise State University in Boise, Idaho, United States. He has been a member of ASCE since 2003. He is now the vice chair of ASCE-GI Geoenvironmental Committee. He has actively conducted research in various fields such as civil, geotechnical, geoenvironmental, and electrical engineering, as well as geophysics. He has published in several journals in various fields from ASCE and ASTM journals to IEEE transactions. He has been nominated for the 2017 ASCE Middlebrooks Award. He has also been the recipient of several research grants and awards from NSF and NASA.

Contents

Preface

This book is based on the contributions of several authors and is an attempt to describe the roles human activities play in causing geohazards. Human activities can directly cause geohazards and result in loss of life and property. Human activities can also indirectly cause geohazards through man-made climate change. This book discusses examples of geohazards, their history, and their formation process. The book emphasizes and differentiates the direct roles and indirect roles—through climate change—human activities can have in causing various geohazards. In the end, the risk of these man-made geohazards and the risk assessment are discussed.

Each chapter keeps the authors' notations that thus vary from chapter to chapter. These authors notations have been maintained to reduce unintended confusion and errors. Readers should be aware of this variation.

Section 1

The topic for Chapter 1 of this book has specifically been selected to showcase the historic extent of geohazard caused by human activities. Chapter 1 studies an ancient case of man-made geohazard that has occurred in the City of Pompeii, a famous ancient city in Southern Italy. Before Pompeii was finally demised by the Plinian eruption in the 79 AD, the city was hit by two alluvial mass flows that damaged the city.

These volcanoclastic deposits emplaced during volcanically quiescent phases of the Somma-Vesuvius volcano were transported and channelized along stream beds. Some of these extended to the immediate proximity of Capua Gate, at the northern side of Pompeii. Human activities played a role when and where an artificial canal was built to supply water to the City of Pompeii. The canal path continued toward Vesuvius Gate and, then, toward the Villa of Mysteries. The first flood deposits, released from hyperconcentrated slumps and debris flows, were not transported through the artificial canal and affected a wide area of the Sarno Plain. The second flood was caused by the canal's limited width and produced severe damage in the archaic city. Later, the third flood event caused severe damage in the northern part of the city. The geological data prove that where human activities are not designed properly, the water, a resource, can in some cases turn into the cause of a geohazard.

Chapter 2 relates to geohazards caused by mining activities. The chapter presents a methodology to analyze geohazards in the form of large-spread landslides caused by mining activities. This methodology is examined in a case study at coal mine "Suvodol" near the town of Bitola in the Republic of Macedonia. Various phases of the landslide from initiation to global instability and reactivation are studied. This landslide geohazard also leads to a secondary geohazard in the form of mass movements of coal at the toe of the landslide. There was a tertiary hazard due to environmental unfriendly gases produced due to the partial self-burning of coal at the toe of the landslide. This shows a series of direct and indirect natural and man-made geohazards caused by human activities. The suggested methodology can serve as an example for possible use in geohazards occurring in coal mining.

Chapter 3 discusses various types of geohazards induced by human activities in the Kingdom of Saudi Arabia (KSA). The chapter identifies main types of human-induced geohazard formations, distribution, causes, and impacts, illustrated through several case studies in the KSA. Examples of these geohazards include recent land subsidence and resulting earth fissures, sinkholes, effect on expansive soils, and flash floods, causing significant life and property loss. The main human activities discussed in this chapter are groundwater extraction, infrastructure development, and agricultural activities. The chapter then provides more details about a human-induced geohazard in the form of earth fissures and sinkhole.

Section 2

The aim of Chapter 4 is to study the relation between rainfall and landslide occurrence in South Korea. Downscaling a regional climate model (RCM) from the global climate model (GCM) based on Intergovernmental Panel on Climate Change (IPCC) A1B scenario was employed in this chapter to develop a model to predict future rainfall. In Chapter 4, to provide a quantitative correlation between rainfall and landslide occurrence, data on rainfall and landslides in Korea in the 2000s were analyzed. The correlation was studied between the occurrence of landslides and daily, accumulated, and maximum hourly intensity of rainfall volume. Thus, high-risk daily rainfall and its duration are then discerned. Thereafter, the annual average rainfall and its increase during 1971 to 2010 due to climate change and the effect on rainfall are studied. The development of downscaling method using GIS and verification using observed data as a method to reduce the uncertainty of future climate change projection are studied at the end of this chapter.

Chapter 5 studies the indirect changes in the scale and impact of geohazards caused by global warming and environmental change, which indirectly relate to human activities. Examples are Typhoon Morakot in 2009 and Tohoku earthquake and resulting tsunami in Japan in 2011. Chapter 5 emphasizes the importance of hazard management to manage risk and fully understand critical scenarios.

The International Federation of Red Cross (IFRC) and Red Crescent Societies have identified technological or man-made hazards as events that are caused by humans and that occur in or close to human settlements. These hazards include environmental degradation, pollution, and accidents. In this chapter, three case studies of accidents caused by man-made geohazards in Taiwan are studied: a highway in southern Taiwan, a freeway in northern Taiwan, and an airport runway in the Taoyuan International Airport. The causes and impacts of these incidents are described to provide lessons about management of man-made hazards.

Section 3

Chapter 6 emphasizes on the damage and risk aspects of geohazards more than the generating process of geohazards. Some of the discussed aspects are the often neglected damage evaluation and oversimplified prediction. This oversimplification results in poor understanding of risks and mistaking them with a mere expression of probability or likelihood of geohazards.

Chapter 6 will use numerous case studies to study technical glossary of risk, damages, crises, multidimensional consequence analysis, and definition of risk tolerance. This chapter also focuses on ethical (geoethical) issues linked to geohazard caused by human activities, mitigation decisions, and possible unintended consequences. This misperception of risks extends to private, public, corporate, and government entities. Geohazards cast a long and often misunderstood shadow on human activities, development, and survival. Hence, understanding how to model geohazard consequences and their risks can help alleviate human and environmental suffering, resulting in sustainable development. The chapter concludes with the root cause of odd human behavior causing risk as a survivor bias.

Dr. Arvin Farid
Civil Engineering, Boise State University,
United States

Section 1

Direct Manmade Geohazards

The Water Supply System of Ancient Pompeii (Southern Italy): From Resource to Geohazard

Maria Rosaria Senatore, Maddalena Falco and Agostino Meo

Additional information is available at the end of the chapter

http://dx.doi.org/10.5772/64413

Abstract

Pompeii, a famous ancient city in the southern Italy, was finally demised by the Plinian eruption in the 79 AD, but, long before it was hit by two alluvial mass flows that damaged the city. These pre-79 AD volcaniclastic deposits had been emplaced by avalanches, slumps, and associated debris flows (secondary lahars) during volcanically quiescent phases of the Somma-Vesuvius volcano. These deposits were transported and channelized along stream beds. Some of these extended to the immediate proximity of Capua Gate, at the northern side of Pompeii, where an artificial canal was built to supply water to the city. The canal path continues toward Vesuvius Gate and then, toward Villa of Mysteries. The flood deposits were released from hyperconcentrated slumps and debris flows. The first flood event, not transported through the artificial canal, took place before the foundation of the city (764 BC) and has affected a wide area of the Sarno Plain. The second one, occurred during the fourth century BC, was caused by the canal's limited width and produced severe damage in the archaic city. Instead, the third flood event occurred in 170 BC and caused severe damage in the northern part of the city. The geological data prove that the water, as resource, in some cases can turn into a geohazard.

Keywords: geological stratigraphy, sedimentology, water supply, artificial canal, flood event, ancient Pompeii

1. Introduction

Pompeii, a famous ancient city in the Southern Italy, is located southeast of Naples in the Sarno Plain at the base of the Somma-Vesuvius volcano and about 2 km from the present Tyrrhenian coastline (**Figure 1**). The Sarno Plain is part of the Campania Plain, a wide Plio-Pleistocene tectonically depressed area (graben) bounded by Mesozoic and Cenozoic carbonate mountains. The graben is partially filled by alluvial, transitional, and marine deposits that are interbedded with pyroclastic deposits mainly from the eruption of the Somma-Vesuvius [1]. The geography and the development of land and population of the Campania Plain have all been conditioned by the volcanic activity [2–4]. The Late Pleistocene and Holocene volcanic activity of the Somma-Vesuvius is characterized by catastrophic Plinian and sub-Plinian eruptions, followed by inter-Plinian and quiescence phases [5–8]. During the settlement of Pompeii, the volcanic activity was weak or absent and the population ignored how danger was the area.

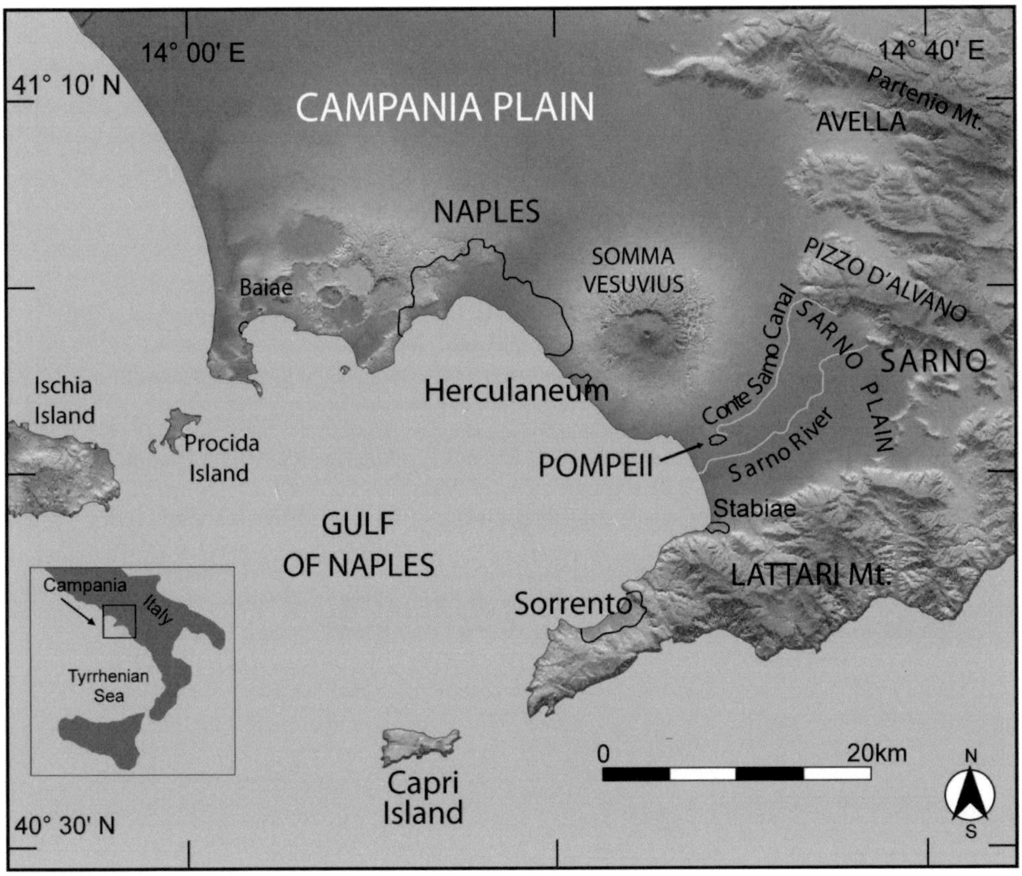

Figure 1. Location map of the Campania Plain – Gulf of Naples. Pompeii, other population centers, and geographic features are showed.

Pompeii was founded at the end of the seventh century BC by the Oscans, a population from central Italy [9, 10]. The town was built on a lava flow or on a separate volcano [11] associated with eruptive events of the Somma-Vesuvius.

Long before its major destruction by the well-documented earthquake in 62 AD [12, 13], and its final demise from the Plinian eruption in 79 AD (called the Pompeii eruption; e.g., [14–17]), Pompeii was damaged by two alluvial mass flows [18, 19]. These pre-79 AD volcaniclastic deposits had been emplaced by avalanches, slumps, and associated debris flows (secondary lahars) during volcanically quiescent phases of the Somma-Vesuvius volcano [20]. These deposits were transported and channelized along stream beds, some of which, extended to the immediate proximity of the northern wall of the city. Nowadays, there are no obvious rivers that would indicate how gravity flows would have reached into the walled city, but there is a stream, named Conte Sarno Canal, extending from the base of the Pizzo D'Alvano Mount (1133 m elevation; **Figure 1**) about 15 km to the northeast from Pompeii. On the northeastern side of the city, the stream shows a large bend (meander) due to the sudden change of the topographic relief occurred as a result of the barrier caused by the lava mound upon which Pompeii was built. The stream originally flowed from the Avella Mountains (**Figure 1**) and, during the Samnite occupation of the city (V–IV century BC), was associated with springs located at the base of the Pizzo D'Alvano ridge [21]. Borehole data collected northwest of the city indicate that a fluvial system reached Pompeii outside of Capua Gate (**Figure 2**). According to [18], the fluvial system was an artificial branch of the Conte Sarno Canal that was diverted

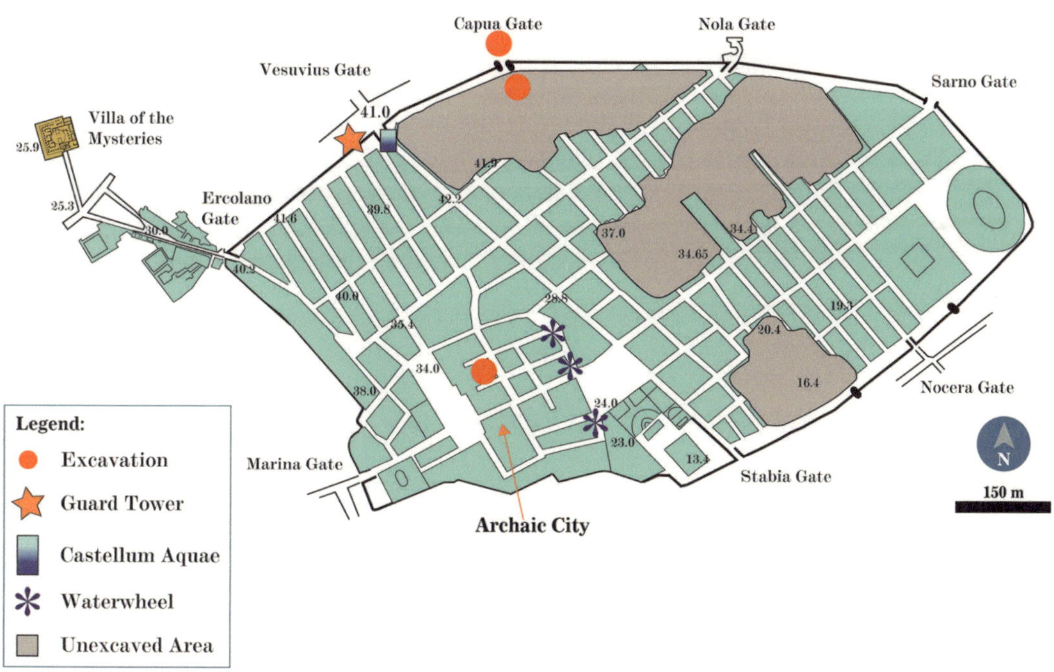

Figure 2. Archaeological area of Pompeii. The archaic part of the city and other archaeological features are showed and the position of archaeological excavations and of unexcavated areas is located. The numbers correspond to the topographic elevation.

toward west and had most likely been excavated to supply the city with water. It was constructed by Samnites as evidenced from the modifications performed along the path of Conte Sarno Canal discovered by [21].

This chapter has two aims: the first is to show the characteristics and the path of the artificial canal discovered to the north of Pompeii, which provided water to the city; the second is to detail the flood units by new borehole data carried out in the south of the ancient city.

Previously the historians studying Pompeii have long suggested that the city's water needs were derived from the Sarnus River (modern Sarno River; **Figure 1**), the largest fluvial system in the area [22–24]. However, nowadays, the modern Sarno channel is positioned to the southeast and south of Pompeii, and the meandering course of the ancient Sarnus River and its delta, identified by analysis of the sediment collected in boreholes, was located at least 1 km south and southwest of the ancient city walls [18, 24–29]. Moreover, the elevation pattern within the city shows that Capua and Vesuvius gates are both positioned at highest elevations (**Figure 2**). Therefore, they occupy strategic points for distribution of the city's water supply. It was from here that water of the artificial canal, entering into the city, discharged by gravity, was able to activate three water wheels (**Figure 2**) located at the edge of the archaic city [30, 31]. However, this artificial canal was also very dangerous because it had been the cause of two of three floods that led to extensive damage to the city. In fact, Senatore et al. [18] have identified, both within the city and outside it, three units referred to debris-flow deposits dated between the eighth and the second century BC. These mass flows are interpreted as having been triggered primarily by intense rains and channelized via the stream that once extended from high reliefs toward Pompeii and, then, through the artificial canal that reached the city. According to these authors, one of these events may have been partially responsible for urban decline during the fourth century BC. New data on the characteristics and distribution of the alluvial deposits related to the two more recent flood events will be analyzed. The interpretation of geological data will prove that a resource, the water, in some cases can turn out to be a geohazard.

2. Water supply system and flood events

2.1. Method

The aim of the researches, carried out in the Pompeii territory since 1995, has been the reconstruction of the paleo-landscape prior to the AD 79 Vesuvius eruption by means of geological stratigraphy and facies analysis. As the studied area is strongly urbanized, about 100 continuous drill-cores were carried out. The detailed stratigraphy of sediments in these drill-cores has been the base reference to re-interpret about 400 logs of older drill-cores. In this chapter, the results of analyses of several boreholes recovered to the northwest (C in **Figure 3**), south, and inside of the city (F in **Figure 3**) are detailed.

Several archeological excavations in the city were analyzed (**Figure 2** and S in **Figure 3**). An electrical resistivity tomography (ERT) profile (TM1 in **Table 1** and **Figure 3**) was recorded on the unexcavated front of a dig carried outside of Capua Gate (**Figure 2**) made by the Japan Institute of Paleontological Studies of Kyoto [32]. The dig brought to light an artificial canal and the TM1 ERT profile analyzed by [18] was made to obtain additional information on the subsurface stratigraphic architecture.

Since 2013, four more ERT profiles were carried out (**Table 1** and **Figure 3**) to reconstruct the path of the artificial canal. The equipment included an MAE A3000E Georesistimeter. The electrical-resistivity measurements recorded were processed through the inversion software RES2DINV by GEOTOMO INTERNATIONAL. The Wenner-Schlumberger and dipole-dipole-array methods were employed as a measure of resistance distribution; Res3DInv software was used for data interpretation. Additional information on the geoelectric equipment and settings used are available in two internal reports [33, 34].

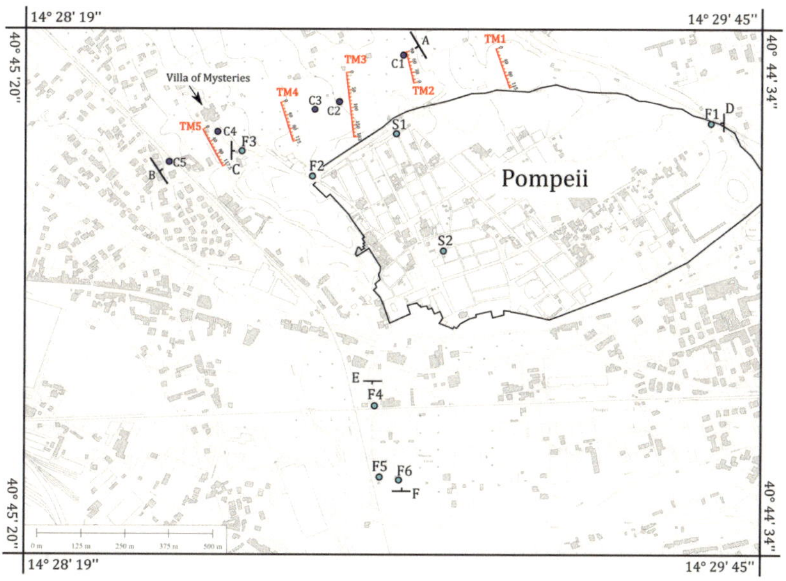

Figure 3. Position of: (C) boreholes passing through the channel units; (F) boreholes passing through flood units; (S) excavations in the city; and (TM) electrical resistivity tomography (ERT) profiles. Section traces are also indicated.

Tomography	Interelectrode spacing	Electrode number	Profile length	Trending	Maximum depth reached
TM1	5.0 m	24	115 m	N 350° E	20.0 m
TM2	1.9 m	48	89.3 m	N 344° E	17.0 m
TM3	2.5 m/5.0 m	72	182.5 m	N 350° E	23.0 m
TM4	5.0 m	24	115.0 m	N 340° E	22.5 m
TM5	2.5 m	48	117.5 m	N 330° E	23.0 m

Table 1. Length, interelectrode spacing, electrode number and maximum depth below the modern topographic surface reached by each TM profile are listed.

Drilling of the cores was performed without the use of circulation fluid to better preserve sedimentary structures, textures, and fabric. Macroscopic characters of the core sediment were defined by a caliper for granules and pebble-size clasts while the grain-size of sand was determined optically by using visual comparison charts. These also allowed to assess clast rounding, sphericity, and sediment sorting. The sediment color was determined by means of the Munsell Soil Color Charts [35], and the thickness of sediment units was defined according to [36]. Selected samples were also analyzed and statistical parameters were even calculated using standard methodologies [37, 38]. Graphic stratigraphic logs were plotted of each drill-core examined.

The sediment cores and logs that constitute the geostratigraphic archive for the study area are stored at the Laboratory of Applied Researches of the Soprintendenza Archeologica at Pompeii.

The AMS radiocarbon analysis reported by Senatore et al. [18] is used to insert the identified units in a chronostratigraphic framework. The base map of **Figures 3** and **7** is an official georeferenced topographic map produced at 1:5000 scale.

The geological interpretations were integrated with the available archeological information.

3. Results

3.1. Stratigraphic units to northwest of Pompeii

The stratigraphic units, identified in the boreholes carried out northwest of Pompeii (**Figure 3**), are composed mostly of volcaniclastic deposits both in primary deposition (eruptive products) and secondary deposition (reworked deposits). Their thickness is from centimeters to several meters, with a highly variable lateral distribution.

Seven stratigraphic units have been identified in a section trending northeast-southwest (from A to B in **Figure 3**). From the topographic surface, they are (**Figure 4**):

- Uc1 represents the deposition following the AD 79 eruption and consists of volcaniclastic sand with brown clay matrix. Plant matter, especially roots, are present. In the upper unit, the sediments are mixed with material linked to the human activity, mainly fragments of brick and pottery. The thickness ranges from few centimeters to 3 m. The basal contact is always sharp.

- Uc2 represents part of the AD 79 eruption deposits and consists of two layers of pumice. The first one is composed of gray pumice, several centimeters in diameter in a volcaniclastic fine sand matrix. The second one is composed of white pumice, few centimeters in diameter. In some cases, the gray and white pumice are mixed to form a single layer. The thickness of the unit is from about 2 m to about 5 m.

- Uc3 represents the Roman and pre-Roman deposits and consists of brown coarse to fine well-rounded volcaniclastic sand. Rounded pumice (few centimeters in diameter) and lapilli clasts, and angular and subangular fragments of artifacts and of animal bones are found in this unit.

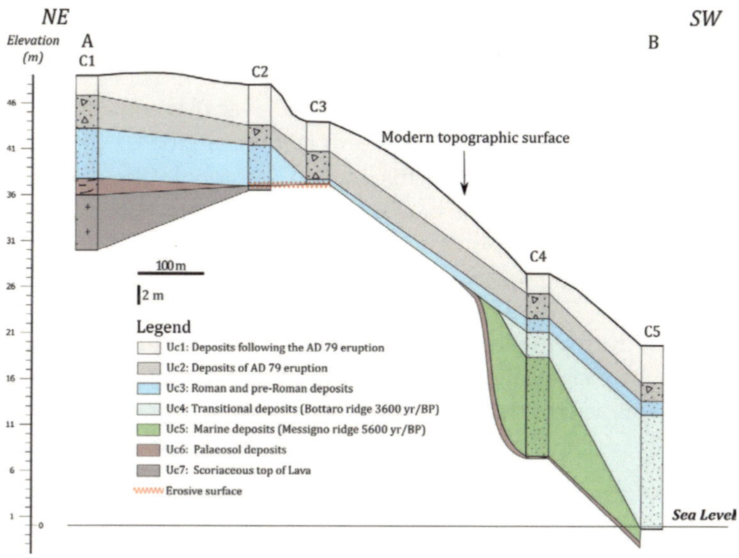

Figure 4. Cross section (A and B) showing the stratigraphic architecture of the units that constitute the northern Pompeii succession (position in **Figure 3**).

The character of the Uc3 deposits allows to define a fluvial channel and the Uc2 deposits as a channel fill while the Uc1 deposits cover the previous units hiding the preexisting morphologies.

- Uc4 is constituted by dark gray, coarse to very fine volcaniclastic deposits with rounded centimetric pumice and lapilli clasts. These deposits are found in the C5 borehole, showing a thickness of about 5 m, and in the C4 borehole with a thickness of 12 m. They are typical of transitional environment and have been correlated to the well-known Bottaro ridge deposits [18], cropping out southwest to the archeological site. They represent an ancient shoreline with radiocarbon age of about 3600 yr/BP [39].

- Uc5 is composed of dark yellow, silty clay deposits with centimetric, rounded, gray pumice and lava clasts. They are found in the C4 borehole with a thickness of about 10 m. The character of the sediment, well known in other analyzed boreholes, allows the correlation to a marine environment linked of the Messigno ridge deposits [18, 23], cropping out southeast to the archeological site inland to the Bottaro ridge. Messigno ridge also represents an ancient shoreline with radiocarbon age of about 5600 yr/BP [39].

The Messigno and Bottaro ridge deposits are found at higher elevations than those with the same age studied in other tectonically stable areas. Significant Holocene ground movements at Somma-Vesuvius area are in fact recorded [40–42].

- Uc6 is constituted by very dark brown, silty clay deposits with weathered white pumice clasts, some millimeter in size, and some remains of roots. This layer is a paleosol and is present at the base of C4 borehole below the Messigno ridge deposits with a thickness of several centimeters, in the C1 borehole at the base of the Uc3 unit, and on the top of the Uc7 unit with a thickness of 2 m. This is lacking in the C2 and C3 boreholes, probably due to an artificial excavation.

- Uc7 is represented by the scoriaceous top of the lava layer that constitutes a morphological high on which the ancient city was built. This unit is found in the C1 borehole where, below the scoriaceous layer, the lava is present; while in the C2 borehole, the scoriaceous layer is just reached. The Uc7 unit is considered the base of the northwestern Pompeii succession.

3.2. Electrical resistivity tomography profiles

Four electrical resistivity tomography (ERT) profiles were acquired to obtain additional information on the subsurface paleogeography based on the water content in the sediment referring to the resistivity values that are from about 10 ohms/m, indicating high humidity up to water presence in the sediment, to 2900 ohms/m, indicating complete absence of water.

The profile trend is NNW-SSE (**Figure 3**). **Table 1** shows the length, interelectrode spacing, electrode number and maximum depth below the modern topographic surface reached by each TM profile.

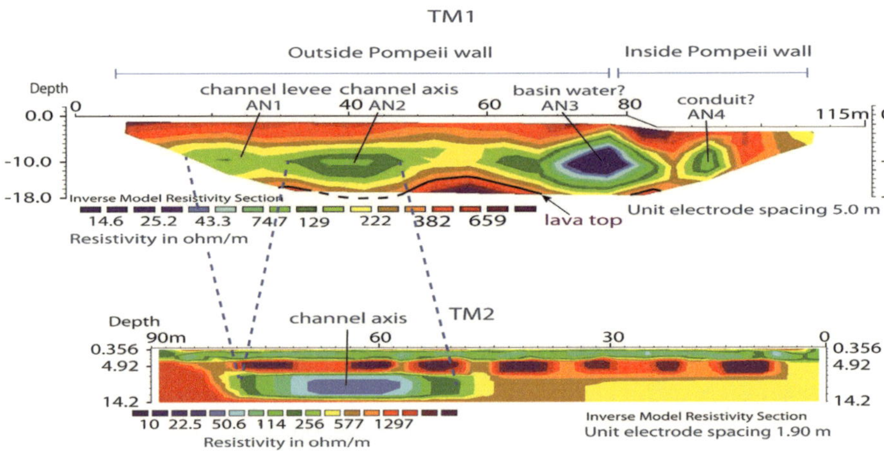

Figure 5. Electrical resistivity tomography (ERT profiles; location in **Figure 3**) showing the characters of the artificial canal (see text for detail) (TM1 modified from [18].

The TM1 profile shows two resistivity anomalies (AN1 and AN2), with resistivity values ranging between 222 ohms/m and 129 ohms/m (**Figure 5**). These anomalies are interpreted, respectively, as the levee and axis of an artificial canal since this profile was performed on the unexcavated dig-front of an archeological excavation outside of Capua Gate (**Figure 2**) and made to examine the subsurface beneath the 79 AD eruption deposits [18]. The archeological excavation has revealed the presence of an artificial canal, which is in the coincidence of the anomaly AN2 on TM1 of **Figure 5**, as there is a close match with regards to both its position relative to electrodes and its depth beneath the present topographic surface. The AN1 on TM1 represents the levee of the canal (see **Figure 6** in [18]). Resistivity values ranging between 382 ohms/m and 659 ohms/m, recorded at the base of profile, are interpreted as the top of the lava layer on which the channel is excavated and Pompeii was built.

Two other ERT anomalies are identified on TM1 (AN3 and AN4 in **Figure 5**) that have generally circular shapes, one of which (AN4) occurs in the archeological area that has not yet been excavated. Anomaly AN3, positioned near the wall, presents a series of concentric resistivity values, which range from 129 ohms/m at the periphery to 14.6 ohms/m at the center of the feature. These values suggest the presence of sediments characterized by high humidity or, possibly, water content. The characteristics of anomaly AN3, the base of which is at the same depth as that of the channel mapped in the excavation, have suggested an anthropogenic structure, probably linked to the water supply distribution to Pompeii [18]. Anomaly AN4, with circular profile and smaller size than AN3, has resistivity values at its center comparable to those of the channel (222 ohms/m and 129 ohms/m). This is interpreted as a smaller channeling feature such as a duct or conduit that was probably related to the city's water distribution system as well [18].

The other four ERT profiles were carried out to trace the path of the artificial canal excavated to carry water to the city, starting from the wide meander of the stream flowing from the inland mountains. In the TM2 profile (**Figure 5**), the canal is identified between electrodes 50 and 76 and between about 7 m and 14 m in depth while the resistivity ranges from 10 ohms/m to 114 ohms/m. In the TM3 profile, the canal is identified between electrodes 70 and 85, and at depth from 5 m to about 20 m. The resistivity ranges from about 50 ohms/m to 114 ohms/m. In these two profiles, the shape of the channel is unnatural, clearly artifact, to allow the flow of the water in the canal by gravity.

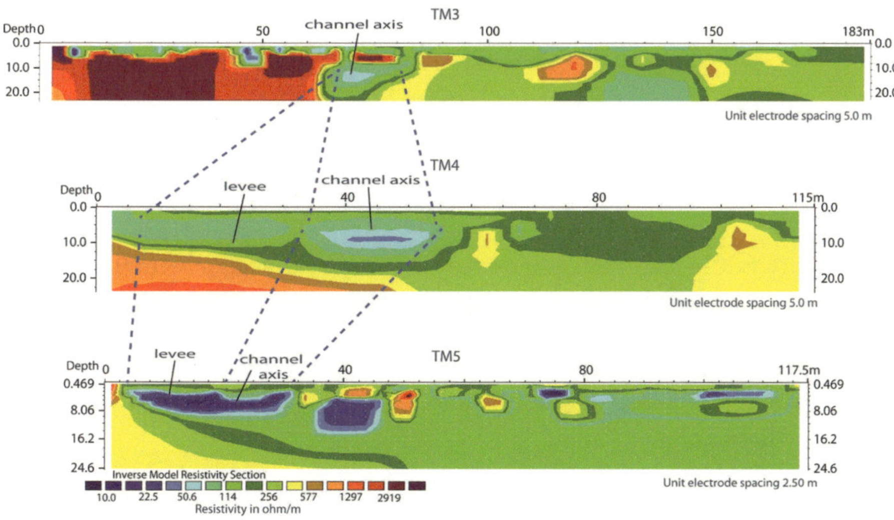

Figure 6. Electrical resistivity tomography (ERT profiles; location in **Figure 3**) showing the characters of the artificial canal (see text for detail).

TM4 and TM5 profiles show the canal between electrodes 35–55 and 25–32, respectively, where the depth is from about 2 m to 10 m (**Figure 6**). The resistivity values are between 10 ohms/m and 114 ohms/m.

Figure 7 shows the path of the canal, which develops from Capua Gate, where, according to [18], a water basin and a conduit, supplied water to Pompeii. The water, entering the city, was then distributed utilizing the gravity. In fact, as stated before, the elevation is greater in this area, and it gradually decreases toward Stabia Gate and the archaic part of the city on the edge of which, the flowing water activated the water wheels (**Figure 2**). The channel path continues toward Vesuvius Gate, touching a farm (Villa Rustica Suburbana) with a foundry [43, 44] and then toward Villa of Mysteries.

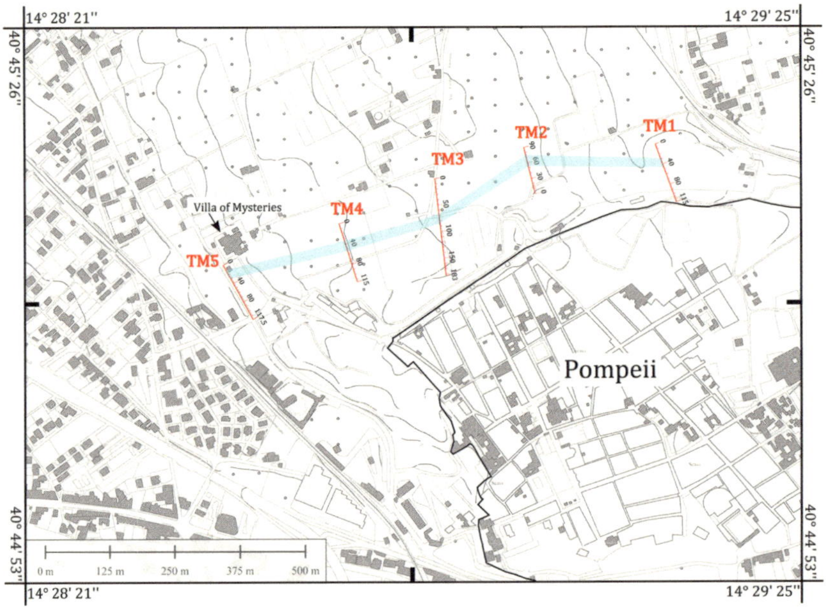

Figure 7. Reconstruction of the path of the artificial canal made by means of ERT profiles and sediments collected in the C boreholes.

The low-resistivity values recorded on the ERT profiles, in connection with the canal, indicate sediment characterized by high humidity up to contain water. They suggest that the canal incision, even today that it is filled by sediments, represents a preferential path for the water flow below the topographic surface.

3.3. Mass gravity flow units

Three flow units, termed Uf1, Uf2, and Uf3 (**Figures 8** and **9**), from the lava base of the succession upward, have been identified in the boreholes carried out the city and the surrounding area (**Figure 3**). Root structures at boundaries between the units indicate that some time has elapsed between the deposition of different mass-flow events.

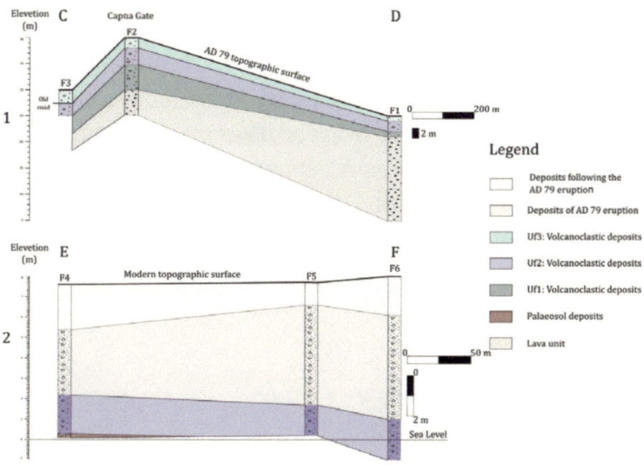

Figure 8. Cross sections: (1) through the flood units (F) to north of Pompeii; (2) through the flood units to south of (position in **Figure 3**). Capua gate is the place where the water enters into the city (position in **Figure 2**).

- Uf1 is composed of massive volcaniclastic deposits with rounded volcanic clasts, rounded to angular fragments of animal bone and plant matter. The unit has a thickness from 1 to 5 m, and rests on the lava upon which Pompeii was built (F1 and F2 in **Figure 8(1)**). The radiocarbon-dated animal bone fragments provided a calibrated age of 764 years BC [18].

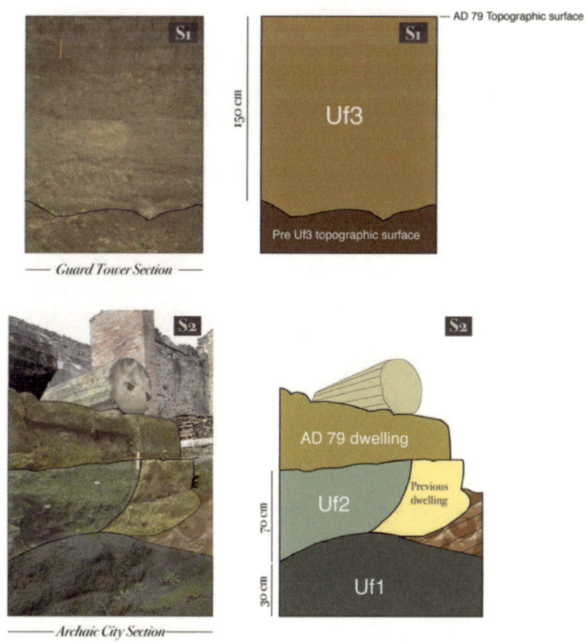

Figure 9. The flood units cropping out in the archaeological excavation and laterally to the Guard Tower door (see text for details).

- Uf2 is identified within and outside city walls (**Figure 8(1 and 2)**), and is constituted of massive volcaniclastic deposits, mainly structureless, or with some thinner cross or planar lamination at the base unit. The matrix is prevalent, with clasts randomly oriented, or, some, may show imbricate structures. The clasts are represented by rounded volcanic clasts and calcareous pebbles, rounded to angular fragments of brick and ceramics, plaster, animal bone, and plant matter. This unit has an average thickness of ~2 m. In the archaic city, the Uf2, between two construction levels, incises Uf1 and covers an older building level (S2 in **Figure 9**). It can be reconstructed that the older dwelling, built upon Uf1, has been damaged by the Uf2 deposits. Subsequently, a new structure was built at a higher elevation upon the Uf2 and was used until its destruction by the 79 AD eruption.

Figure 10. Guard Tower (position in **Figure 2**): the door is below the AD 79 topographic surface and on its left the Uf3 section crops out (see S1 in **Figure 8**).

- Uf3 is composed of matrix-prevalent volcaniclastic deposits with randomly distributed clasts that commonly comprise rounded volcanic clasts and rounded to angular fragments of pottery, plaster, animal bone, and plant matter. These deposits have an average thickness of ~1 m, occurring along the northern city wall (F1 and F2 in **Figure 8(1)**). In F3, Uf3 buries an ancient road trending from the city toward Villa of the Mysteries (**Figures 3** and **8(1)**). Section S1 (in **Figure 9**) is located laterally to the door of a Guard Tower (**Figures 2** and **10**) and shows the Uf3 character. The Guard Towers were added to the city wall during the second century BC

[45]; the tower doors, openings at their base, nowadays, occur beneath the topographic surface of the AD 79 (**Figure 10**). They have been buried by Uf3 deposits and were used until the AD 79 eruption after the removal of the Uf3 material. The radiocarbon-dated animal bone gave a calibrated age of 170 years BC [18].

4. Conclusive remarks

The sediment characteristics of the three Uf units indicate that mass gravity mechanisms, especially debris flow, were the dominant processes responsible for their transport, and the two younger units had flooded Pompeii causing severe damage to the city. The volcaniclastic sediment with matrix-supported clasts likely originated as slope collapse and avalanche displacement from the flanks of calcareous terrains of mountains to the NE (Pizzo D'Alvano area; **Figure 1**). During landslides, slumped masses of unconsolidated material can be transformed to high-concentration debris flows as has been recorded in volcanic areas elsewhere [46]. Confined within downslope-trending depressions, such as channels, flows can travel considerable distances toward lowlands by expanding in volume during transport through a bulking mechanism that involves incorporation of additional sediment and water [47]. In the studied area, these deposits were released from hyperconcentrated slumps and debris flows that had incorporated sediment and water during the course of downslope transport in the fluvial channel. The first flood event, which had not occurred through the canal, took place in 764 BC, before the foundation of the city was built [18], and has affected a wide area of the Sarno Plain.

The available data allow to reconstruct the hypothetical phenomena that can be occurred in a temporal sequence during the emplacement of the second and third flood events, linked both to the canal built by the Samnitic population for water supply [18]. Therefore, the flow, in the fluvial channel, reached the great bend to north of Pompeii. Hence it was channeled in the artificial canal and continued its course within it. In the proximity of the Capua Gate, the canal width being narrower than that of the fluvial channel, the flow overflowed its banks thus flooding the city. This event caused severe damage in the archaic city (S2 in **Figure 9**). According to [18], this flood could have occurred during the fourth century BC. The third flood event, which took place in 170 BC [18], whose sediments were found only in F1, F2, and F3 boreholes (**Figure 8(1)**) and in S1 section (**Figure 9**), seems to have caused severe damage only in the northern part of the city.

At the Capua Gate inside the city walls, a duct was discovered under the first floor of a building [32]. This feature that was filled with sediment of the Uf3 unit may represent the extension of the duct highlighted by the anomaly AN4 in the TM1 ERT profile [18]. The sediment of the Uf3 unit was also piled against the entrance door of the building discovered at Capua Gate. According to [18], it is proven that at the time of the AD 79 eruption, the building and the duct below the floor were no longer used. It seems that after the mass-gravity flood event that had deposited the Uf3 unit, the water distribution system at the Capua Gate had to be abandoned due to its danger for the city. Hence, a new water supply system had to be organized. In fact, in 80 BC, a circular water basin was built close to the Vesuvius Gate and was connected to an aqueduct originating from the mountains northeast of the town named Avella (Avella Aqueduct [48]). The circular basin was afterwards covered (Castellum aquae, **Figure 3**) and was connected to the new Serino Aqueduct, in 20 BC [49–51]. This last water system was in use until the final demise of the city because of the Vesuvius eruption.

In conclusion the geological data prove that the first system for water supply caused floods that, in turn, caused severe damage to the city. Hence, the water, usually as resource, in some cases can turn into geohazard.

Acknowledgements

Funding was provided by Ministero Università e Ricerca Scientifica (Pon Project 12232; MRS) and Università degli Studi del Sannio (FRA Projects; MRS). MRS is grateful to the staff of the Laboratory of Applied Researches of the Soprintendenza Archeologica di Pompei: Luigi Buffone, Antonio Stanpone, and Vincenzo Di Martino. In memory of Annamaria Ciarallo, the first director of the Laboratory of Applied Researches.

Author details

Maria Rosaria Senatore*, Maddalena Falco and Agostino Meo

*Address all correspondence to: senatore@unisannio.it

Department of Science and Technology, University of Sannio, Benevento, Italy

References

[1] Brocchini D, Principe C, Castratori D, Laurenzi MA, Gorla L. Quaternary evolution of the southern sector of the Campanian Plain and early Somma-Vesuvius activity: insights from the Trecase 1 well. Mineralogy and Petrology 2001;73:67–91.

[2] Luongo G, Perrotta A, Scarpati C, De Carolis E, Patricelli G, Ciarallo A. Impact of the AD 79 explosive eruption on Pompeii, II. Causes of death of the inhabitants inferred by stratigraphic analysis and areal distribution of the human casualties. Journal of Volcanology and Geothermal Research 2003;126:169–200.

[3] Luongo G, Perrotta A, Scarpati C. Impact of the AD 79 explosive eruption on Pompeii, I. Relations amongst the depositional mechanisms of the pyroclastic products, the framework of the buildings and the associated destructive events. Journal of Volcanology and Geothermal Research 2003;126:201–223.

[4] De Simone GF, Perrotta A, Scarpati C. L'eruzione del 472 d.C. ed il suo impatto su alcuni siti alle falde del Vesuvio. Rivista di Studi Pompeiani. 2011;XXII:61–71.

[5] Arnò V, Principe C, Rosi M, Santacroce R, Sbrana A, Sheridan MF. Somma-Vesuvius eruptive history. CNR, Quaderni della Ricerca Scientifica. 1987;13:53–103.

[6] Andronico D, Calderoni G, Cioni R, Sbrana A, Sulpizio R, Santacroce R. Geological map of Somma Vesuvius volcano. Periodico di Mineralogia. 1995;64:77–78.

[7] Rolandi G, Petrosino P, McGeehin J. The interplinian activity at Somma-Vesuvius in the last 3500 years. Journal of Volcanology and Geothermal Research 1998;82:19–52.

[8] Di Vito MA, de Vita S, Piochi M. Il Somma Vesuvio: storia eruttiva e impatto delle sue eruzioni sul territorio. Miscellanea INGV. 2013;18:14–21.

[9] De Caro S. Lo sviluppo urbanistico di Pompei. Atti della Società della Magna Grecia. 1992;1:69–90.

[10] Guzzo PG, d'Ambrosio A. Pompei. L'Erma di Bretschneider, Naples, Electa. 1998:1–160.

[11] Cinque A, Irollo G. Il vulcano di Pompei: nuovi dati geomorfologici e stratigrafici. Il Quaternario, Italian Journal of Quaternary Sciences. 2004;17(1):101–116.

[12] Jacobelli L. I terremoti fra il 62 e il 79 d.C. nell'area Vesuviana: le ragioni di un convegno. In Archäologie und Seismologie, La regione vesuviana dal 62 al 79 d.C. Problemi archeologici e sismologici, Colloquium, Boscoreale 26–27 November 1993. 1995:17–21.

[13] De Simone A. Terremoti precedenti l'eruzione. Nuove attestazioni da recenti scavi. In Archäologie und Seismologie, La regione vesuviana dal 62 al 79 d.C. Problemi archeologici e sismologici, Colloquium, Boscoreale 26–27 November 1993. 1995:37–43.

[14] Lirer L, Pescatore T, Booth B, Walker GPL. Two plinian pumice-fall deposits from Somma-Vesuvius, Italy. Geological Society of America Bulletin. 1973;84:759–772.

[15] Sigurdsson H, Carey S, Cornell W, Pescatore T. The eruption of Vesuvius in A.D. 79. National Geographic Research and Exploration. 1985;1:332–387.

[16] Cioni R, Marianelli P, Sbrana A. Dynamics of the A.D. 79 eruption: stratigraphic sedimentological and geochemical data on the successions from the Somma–Vesuvius southern and eastern sectors. Acta Vulcanologica. 1992;2:109–124.

[17] Cioni R, Civetta L, Marianelli P, Métrich N, Santacroce R, Sbrana A. Compositional layering and syneruptive mixing of a periodically refilled shallow magma chamber: the AD 79 Plinian eruption of Vesuvius. Journal of Petrology. 1995;36(3):739–776.

[18] Senatore MR, Ciarallo A, Stanley J. Pompeii damaged by volcaniclastic debris flows triggered centuries prior to the 79 A.D. Vesuvius eruption. Geoarcheology. 2014;29:1–15. DOI: 10.1002/gea.21458.

[19] Ciarallo A, Senatore MR, Stanley J. Il territorio vesuviano nel 79 d.C. In: Vincenzina Castiglione Morelli, Ernesto De Carolis, Claudio Rodolfo Salerno, editors. Caio Giulio Polibio - Storie di un cittadino pompeiano. Edistampa; 2015. p. 391–405.

[20] Andronico D, Cioni R. Contrasting styles of Mount Vesuvius activity in the period between the Avellino and Pompeii Plinian eruptions, and some implications for assessment of future hazards. Bulletin of Volcanology 2002;64:372–391.

[21] Murano D. Pompei. Donde venivano le acque potabili ai castelli acquari. Napoli, Tipografia Cav. A. Morano and E. Veraldi. 1894:1–147.

[22] Maiuri A. Pompeii. Scientific American 1958;198:68–78.

[23] Cinque A, Russo F. La linea di costa del 79 d.C. fra Oplonti e Stabiae nel quadro dell'evoluzione olocenica della Piana del Sarno (Campania). Bollettino della Societa Geologica Italiana. 1986;105:111–121.

[24] Senatore MR. Pompei, una storia di acqua e di fuoco. In: L'Opinione di. Associazione Ambiente e cultura Mediterranea; 2015.

[25] Pescatore T, Senatore MR, Capretto G, Lerro G, Patricelli G. Ricostruzione paleogeografia delle aree circostanti l'antica città di Pompei (Campania, Italia) al tempo dell'eruzione del Vesuvio del 79 d.C. Bollettino della Societa Geologica Italiana. 1999;118:243–254.

[26] Pescatore T, Senatore MR, Capretto G, Lerro G. Holocene coastal environments near Pompeii before the A.D. 79 eruption of Mount Vesuvius, Italy. Quaternary Research. 2001;55:77–85.

[27] Ciarallo A, Pescatore T, Senatore MR. Su di un antico corso d'acqua a nord di Pompei. Dati preliminari. Rivista di Studi Pompeiani, L'Erma di Bretsheneider. 2003;14:274–283.

[28] Vogel S, Marker M. Reconstructing the Roman topography and environmental features of the Sarno River Plain (Italy) before the AD 79 eruption of Somma-Vesuvius. Geomorphology 2010;115:65–77.

[29] Ciarallo A, De Carolis E, Senatore MR. Water supply and water circulation in ancient Pompeii: resource management and catastrophic events in the past as in the present. Rendiconti Online, Società Geologica Italiana. 2012;21:738–740.

[30] Maiuri A. Pozzi e condutture d'acqua nell'antica città. Scoperta di un antico pozzo presso "Porta Vesuvio". Notizie degli Scavi di Antichità, Accademia Nazionale dei Lincei, Roma. 1931:546–576.

[31] Oleson JP. Water-lifting devices at Herculaneum and Pompeii in the context of Roman technology. N. de Haan and G.C.M. Jansen (Eds), Cura Aquarum in Campania,

Proceedings of the Ninth International Congress on the History of Water Management and Hydraulic Engineering in the Mediterranean Region Pompeii; 1–8 October 1994. Leuven, Belgium: Peeters, 1994:67–75.

[32] Sakai S. La storia sotto il suolo del 79 d.C. Considerazioni sui dati provenienti dalle attività archeologiche svolte sulle fortificazioni di Pompei. Opuscula Pompeiana. 2000-2001;10:87–100.

[33] GTGeotesting s.r.l. Esecuzione di profili di tomografia geoelettrica presso il sito archeologico di Pompei (NA). Relazione di sintesi delle indagini eseguite. 2013:1–30.

[34] GTGeotesting s.r.l. Esecuzione di profili di tomografia geoelettrica e sondaggi geognostici presso il sito archeologico di Pompei (NA). Relazione di sintesi delle indagini eseguite. 2014:1–16.

[35] Munsell A. Soil Colour Charts. Macbeth Division of Kallmorgen Corporation, Baltimore, Maryland 21218. 1975.

[36] Campbell CV. Lamina, laminaset, bed and bedset. Sedimentology 1967;8:7–26.

[37] Folk RL. Petrology of Sedimentary Rocks. University of Texas Publication. 1968:1–170.

[38] Folk RL, Ward WC. Brazos river bar: a study in the significance of grain size parameters. Journal of Sedimentary Petrology 1957;27:3–26.

[39] Barra D. La piana del fiume Sarno [PhD thesis]. Studio del Pleistocene Superiore-Olocene delle aree vulcaniche campane: 1991. 34–59.

[40] Marturano A. Sources of ground movement at Vesuvius before the AD 79 eruption: evidence from contemporary accounts and archaeological studies. Journal of Volcanology and Geothermal Research 2008;177:959–970.

[41] Marturano A, Aiello G, Barra D, Fedele L, Grifa C, Morra V, Berg R, Varone A. Evidence for Holocenic uplift at Somma-Vesuvius. Journal of Volcanology and Geothermal Research. 2009;184:451–461.

[42] Keenan-Jones D. Somma-Vesuvian ground movement and the water supply of Pompeii and the Bay of Naples. American Journal of Archaeology 2015;119:191–215.

[43] Stefani G. Pompei. Vecchi scavi sconosciuti: la villa rinvenuta dal marchese Giovanni Imperiali in località Civita (1907–1908). "L'ERMA" di Bretschneider. 1994:118.

[44] Moormann EM. Villas surrounding Pompeii and Herculanum. The World of Pompeii. 2007:435–454.

[45] d'Ambrosio A. Mura di cinta. Pompei, gli scavi dal 1748 al 1860. 2002:92–93.

[46] Scott KM, Macias JL, Naranjo JA, Rodrigues S, McGeehin JP. Catastrophic debris flows transformed from landslides in volcanic terrains: mobility, hazard assessment, and mitigation strategies. U.S. Geological Survey Professional Paper 1630. 2001:1–59.

[47] Scott KM, Vallance JW, Kerle N, Macias JL, Strauch JL, Devoli G. Catastrophic precipitation-triggered lahar at Casita volcano, Nicaragua: occurrence, bulking and transformation. Earth Surface Processes and Landforms 2005;30:59–79.

[48] Ohlig C. De aquis Pompeiorum. Das Castellum Aquae in Pompeji: Herkunft zuleitung, Verteilung Wassers. J.A.K.E De Waele and E.E. Moormann. 2001:1–483.

[49] Nappo SC. L'impianto idrico di Pompei. Nuovi dati. In: N. de Haan and G.C.M Jansen, editor. Cura Aquarum in Campania, Proceedings of the Ninth International Congress on the History of Water Management and Hydraulic Engineering in the Mediterranean Region Pompeii; 1–8 October 1994; Leuven, Belgium: Peeters; 1994. p. 37–45.

[50] Potenza U. Gli acquedotti romani di Serino. Azienda Municipalizzata Acquedotto di Napoli (AMAN). 2001:1–22.

[51] Matsui S, Sorrentino L, Sakai S, Shimizu Y, Iorio V. La provenienza dell'acqua potabile nell'antica Pompei: un'ipotesi basata sull'analisi chimica dei residui calcarei degli impianti idrici. Documents & Research. 2009:162.

Chapter 2

Geohazards at Surface Coal Mines Caused by Mining Activities

Milorad Jovanovski and Igor Peshevski

Additional information is available at the end of the chapter

http://dx.doi.org/10.5772/66140

Abstract

This chapter presents a methodology of geohazard analyses caused by mining activities at coal mine "Suvodol" near the town Bitola in the Republic of Macedonia. The problems discussed here are connected with landslide with enormous dimensions. The process of sliding happened in several phases, with initial signs of sliding in 1993. The moment of global instability happened on October 27, 1995. Until now, several phases of reactivation are known. Its volume is about 30,000,000 m^3. As a result of mass movements, about 8,000,000 tons of coal is concentrated (blocked) at the toe of the landslide. Upper of the main scarp, spaced about 250 m, the earth-fill dam with a length of about 1000 m exists. The groundwater artesian effects are also present. At the toe of the landslide, the coal is partially involved in a process of self-burning, and it produces environmental unfriendly gases. All these aspects show a very specific combination of natural and man-made hazards that control the stability of the excavation and environment. The specific approach used to define risk scenarios for is then shown briefly. The suggested methodology can serve as an example for possible use in some other problems in coal mines.

Keywords: coal mine, coal self-burning, geohazards, groundwater, landslide, environment, risk, stability

1. Introduction

It is well-known that efficient designing of engineering activities and safe exploitation in coal mines is not possible without knowing in detail set of geological, geotechnical, and groundwater conditions. The main principle is that technology of excavation should always be carefully adapted to the properties of the natural environment and surroundings. This is of special importance having in mind that during the excavation process, there are possibilities for development of important induced geohazard. The induced geohazards can be connected with changes of groundwater conditions, stability of the excavation zone, possible settlements because of dewatering, possible development of coal self-burning process, influence on the surrounding structures, air and groundwater pollution, etc.

This statement is especially emphasized in cases when the exploitation is close to other infrastructure and engineering structures, as a case for coal mine "Suvodol" placed on southwest (SW) part of the Republic of Macedonia (**Figure 1**).

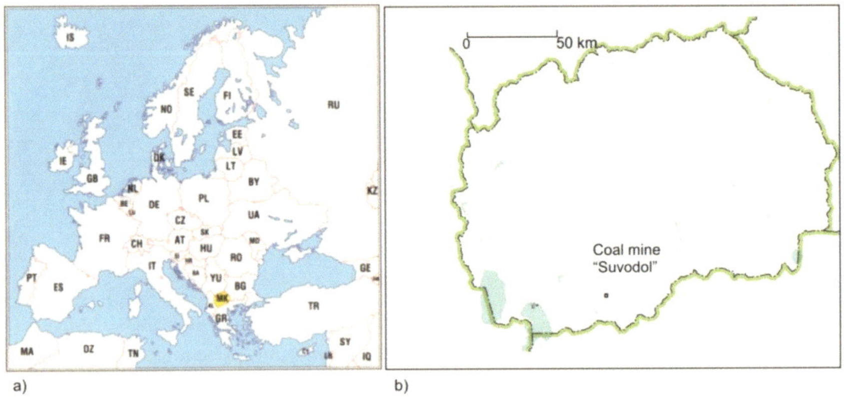

Figure 1. (a) Key map illustrating position of R. Macedonia in Europe; (b) position of coal mine "Suvodol".

The coal mine is a main source for thermal electricity plants with coal production of about 6,500,000 tons per year. More precisely, the analyzed problem in this chapter is related to the northeast (NE) part of the mine, where during longer time large landslide appeared and caused lots of difficulties in the normal work of the exploitation systems [1]. It is also a potential danger for the upstream earth dam, which is spaced about 250 m from the main scarp of the landslide.

To overcome this problem, the authors were involved in several phases of landslide investigations and design phases. The investigations were complex and with large quantity, in order to prepare data for physical and analytical modelling [2, 3]. Later, the data are used as a base for stability and dewatering analyses, protection from surface and groundwater, excavation conditions and so on. The methods are presented within the wider context of an approach to integrate all the relevant information in a similar way as it is given in rock engineering design and construction. Namely, the methodology of developing the so-called rock engineering systems (RES) is firstly introduced in [4]. Here, we will present the used approach in developing geotechnical engineering systems (GES). The entire concept providing overall coherency in approaching engineering problems at coal mines, where the need to study the interactions has always been present.

The key question here is to have correctly carried out investigations of the groundwater and stability conditions at the zone of interaction between the natural environment and the engineering activities, estimations of risks, etc.

http://dx.doi.org/10.5772/66140

A framework for this concept has earlier been given in [3], while the methodology and results are explained in references [2, 3, 5–8].

2. Geological, hydrogeological, and geotechnical conditions of the analyzed area

The coal mass and the unproductive layers at coal mine "Suvodol" have been formed with a process of sedimentation in lake conditions during upper Pliocene. The geological composition is presented with the so-called bottom-coal series with layers of silty sands, productive series of coal and coal-like clay, and layers on the upper part of coal of volcanic material (the so-called trepel). The area of mine is investigated, with very detail and using complex investigation methods in a several phases before opening of the mine, but also during the phase of exploitation. These investigations have been made in the sense of solving the entire geological, geotechnical, and hydrogeological situation on the terrain. For instance, mapping of the wider area, investigation drillings, installing of group piezometers, investigations of the chemical composition of groundwater's, field investigation of filtration coefficient, as well as laboratory analyses of physical and mechanical properties are applied.

To illustrate geological and hydrogeological conditions, some results are presented in **Figures 2** and **3**.

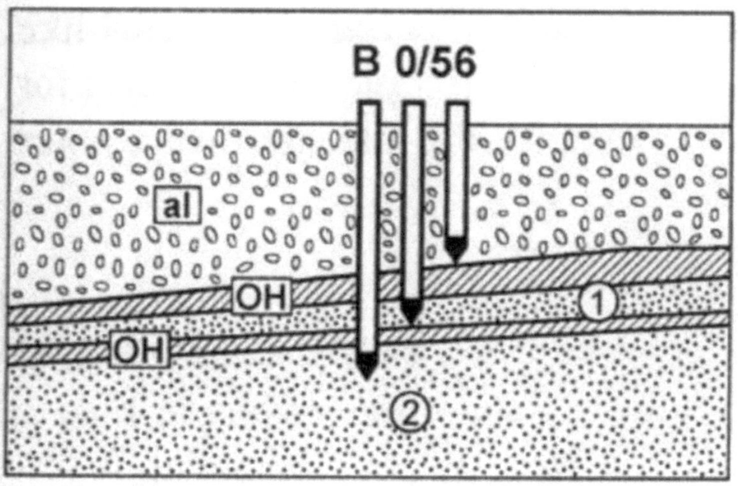

Figure 2. Schematic of installation of triple piezometer in a borehole B 0/56: al – alluvial sediments; OH – coal-like clay; 1 – interstratified aquifer zone; 2 – aquifer zone at the bottom of clay.

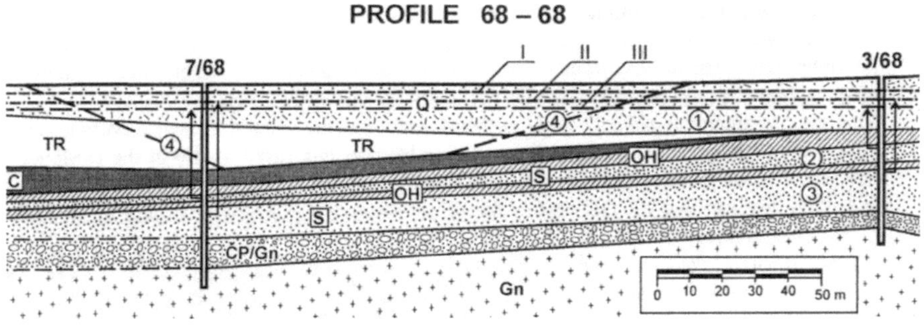

Figure 3. Detail of geological and hydrogeological composition at one zone of coal mine "Suvodol": 1 – Aquifer zone with phreatic line; 2 – Interstratified aquifer zone under pressure; 3 – Aquifer zone at the bottom of the coal layer; 4 – Designed cut; Q – Quarterian silty sand layer; TR – trepel (aquiclude); C – coal; OH – coal-like clay (aquiclude); S – silty sands (aquifer); Gn – gneiss; I – free water table; II – piezometric level for the aquifer zone at the bottom of the coal layer; III – piezometric level for the interstratified aquifer zone under pressure.

Some of the zones at the coal mine are with high lithological heterogeneity, which is the reason why there is heterogeneity of hydrogeological and geotechnical characteristics. By the help of installed piezometers, the presence of several physically separated aquifer zones is shown in **Figure 4**.

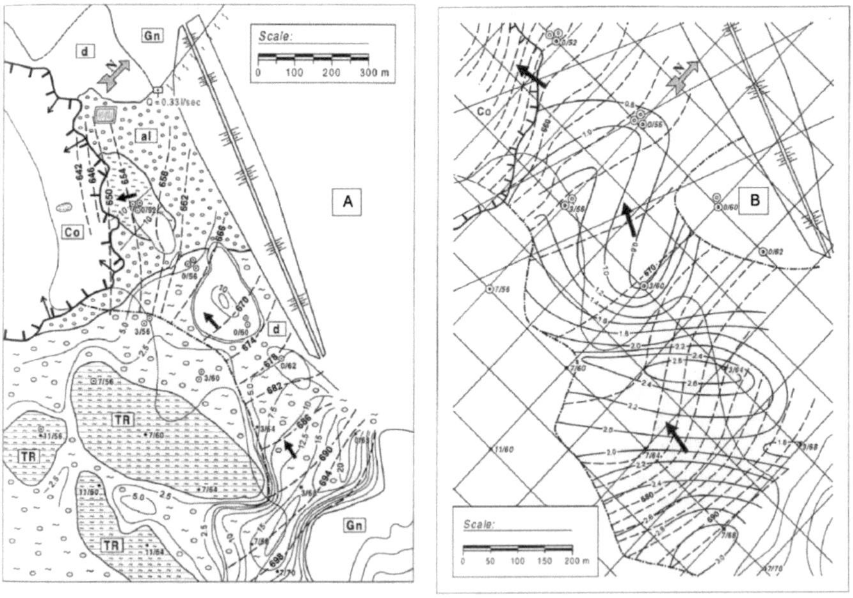

Figure 4. (**A**) Engineering geological map of the NE part of coal mine: ➡ – groundwater flow paths for the aquifer zones; --- 670 --- contour lines of groundwater level; al – alluvial sediments; dl – deluvial sediments; TR – trepel; Gn – gneiss; (**B**) Model of the groundwater movement for the interstratified aquifer zone under pressure at the NE part of the coal mine: ------- 2 -------- the contour line of equal artesian pressure (in bars); Co-colluvial material (active landslide) [5].

Figure 4A presents a model of groundwater movement for the aquifer zone with free-water level. **Figure 4B** presents the model of groundwater flow for the so-called interstratified aquifer zone under pressure, placed between two layers of coal-like clay. The aquifer zone under artesian conditions exist also bellow main coal layer with high values of pressures, affecting stability of the area.

Chemical composition of groundwater is also very important, because it influenced the installed equipment (pumps) for dewatering (**Table 1**).

Content of ions in mg/l	pH	Ca^{2+}	Mg^{2+}	Fe^{2+}	Cl^-	SO_4^{2-}	HCO_3^-	Free CO^2	Rest
Aquifer zone with free water table	6.8	20.1	12.5	0.4	158	194	701.5	–	15.5
Interstratified aquifer zone under pressure	6.5	216	21.8	2.6	184	256	760.5	70	6.8
Aquifer zone at the bottom of the coal layer	5.7	140	24.3	4.8	19	43.6	549.3	111	1.3

Table 1. Typical chemical composition of the aquifer zones.

It can be noticed that there are aggressive groundwater components with the presence of gas (CO_2, Radon, and others), which is important from ecological aspect and working conditions at the mine.

3. Brief overview of landslide elements

The complex geological and hydrogeological elements, combined with excavation for coal production, were a reason for occurrence of large landslide on the NE part of the mine. The initial phase of activation was at the end of 1995, but several large reactivation phases were also present in 1997 and 1998. Some smaller movements were also present in parts of the landslide continuously till present days. In order to illustrate the scale of the event, the main elements of the landslide are given in **Table 2**.

Landslide element	Value
Length (m)	About 1700
Width (m)	Min. 650–Max. 880
Area (m^2)	About 1,050,000
Volume (m^3)	About 30,000,000
Depth to sliding zone (m)	Min. 14–Max. 55

Table 2. Main landslide elements.

The most important characteristics for the main kinds of sediments are the high plasticity of coal-like clay and Pliocene silts, high value of the coefficient of uniformity $C_{u,}$ and low shear strength of coal-like clay and silts with high plasticity. The typical granulometric curves and the plasticity chart of clay are given in **Figure 5**.

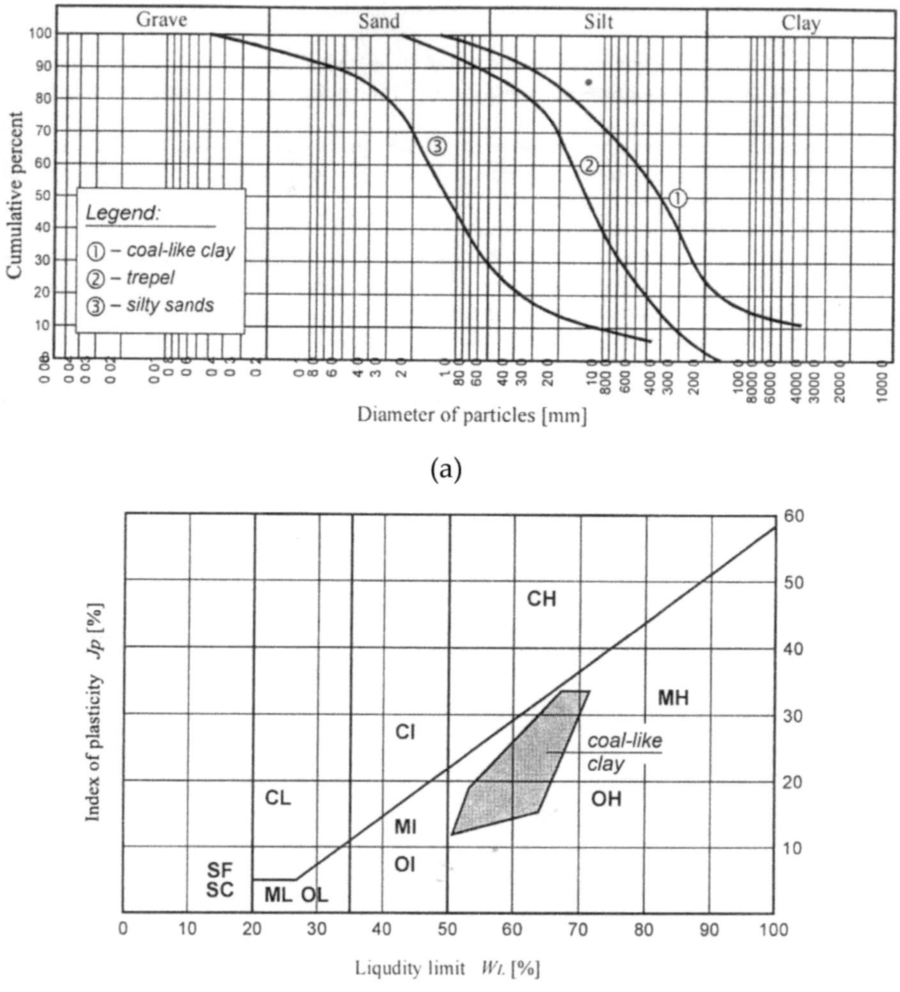

Figure 5. (a) Typical granulometric curves of the most characteristic sediments; (b) plasticity chart for coal-like clay.

Graphical presentation of main landslide elements is presented in **Figures 6** and 7. From **Figures 6** and 7, it is obvious that the main lithological units are very disturbed and displaced from their original position. Fortunately, during the process of sliding, the retrogressive extension of movement stops about 250 m from the earth-fill dam and during the main phase of activation, there were not injured working stuff.

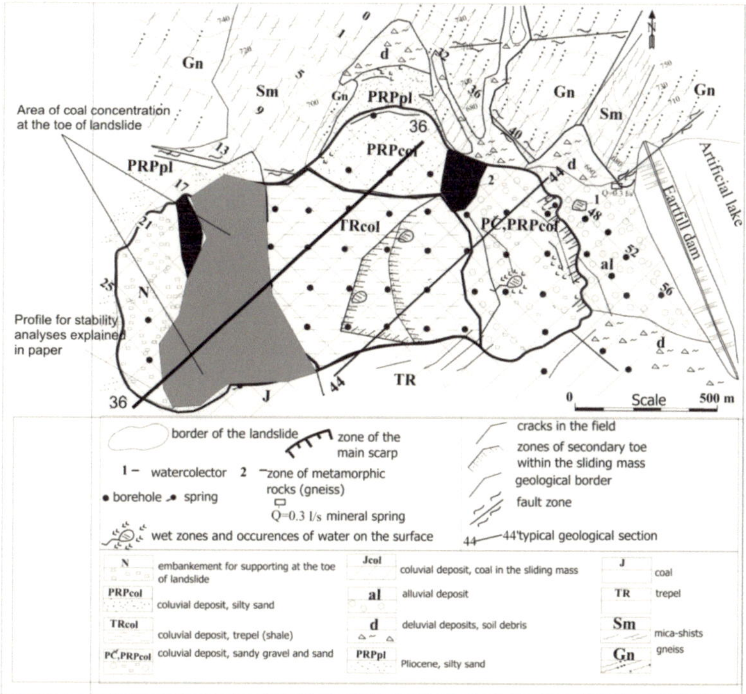

Figure 6. Simplified engineering geological map of the landslide in relation to the earth-fill dam [2].

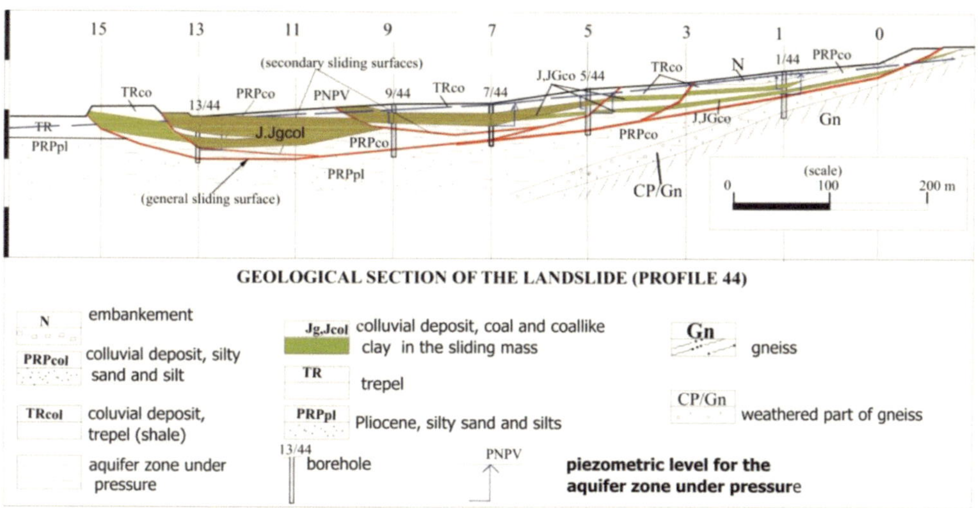

Figure 7. Presentation of geological composition of the landslide along one profile.

Results from the investigations also indicate that the sliding surfaces are very deep, usually along coal-like clay and silts with high plasticity (**Figure 7**).

To illustrate this, we present the map with relative subsidence and uprising of the field, after the phase of main activation and the map of the thickness of the landslide (**Figure 8**) [2].

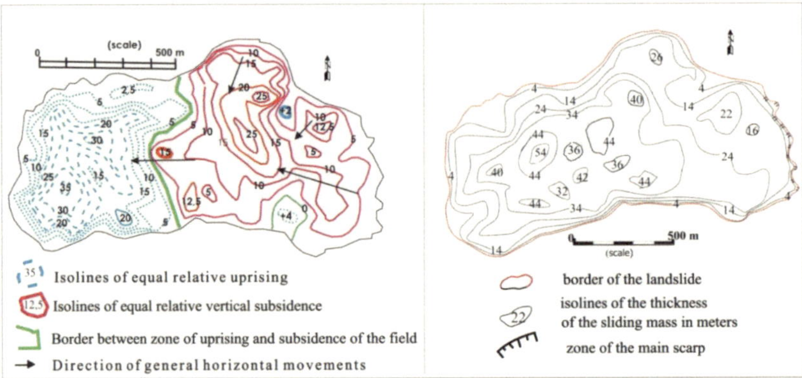

Figure 8. Isolines of relative vertical uprising and settlements in meters (left side) and isolines of landslide body thickness (right side).

It is more than obvious that the range of horizontal and vertical displacements is very large, and the thickness of the landslide body is very high. In one word, the event can be described as a "small tectonic."

A lot of secondary scarps and zone of "secondary toe" were also defined. Artesian effects are directly observed during drilling. A huge quantity of sand was transported from drilling bottom to the ground surface, because of high artesian pressures and hydraulic gradients. Analyzing all data, it can be noted that groundwater conditions have the greatest influence on the stability. The aquifer zone under artesian conditions with gases is especially important. Another important hazard and very restrictive factor was the process of self-burning, which happened because of coal's direct exposition on the fresh air.

Shortly, the problem is too complex and unique that every technical action is always connected with numerous restrictions and risks.

4. Methodology of hazard and risk analyses

Analyzing the behavior of the landslide from the time of its occurrence until present days, some facts can be underlined such as follows.

- After the main movements, the initial technical measures are applied as unloading and crack filling in critical zones, in order to minimize any further retrogressive development of sliding in the dam direction,
- The excavation of the coal was stopped at this area,

- The toe of the landslide was supported with embankment zone (see **Figure 6**),
- The hydrostatic influences of the aquifer zones for the zone between the earth-fill dam and the main scarp was decreased with dewatering wells,
- The entire surface of the landslide was graded and drained for fast atmospheric water influences, etc.

In addition, there were several phases of smaller landslide reactivations. An important new element and very restrictive additional factor after the sliding was a process of self-burning, which happened because of coal's direct exposition on the fresh air. Thus, to minimize these effects, two main risk scenarios are analyzed (**Figures 9** and **10**).

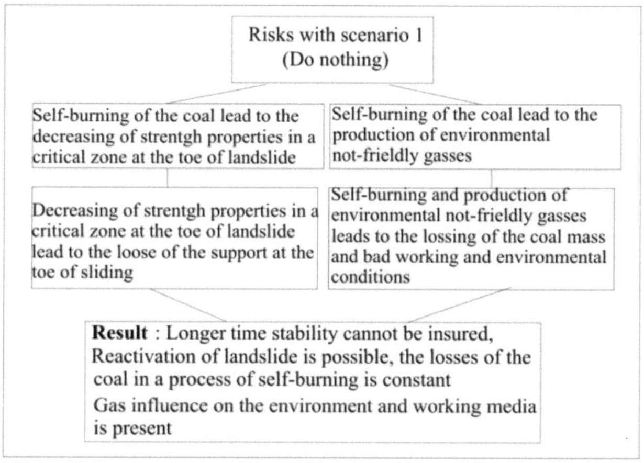

Figure 9. Presentation of main problems in Risk Scenario 1.

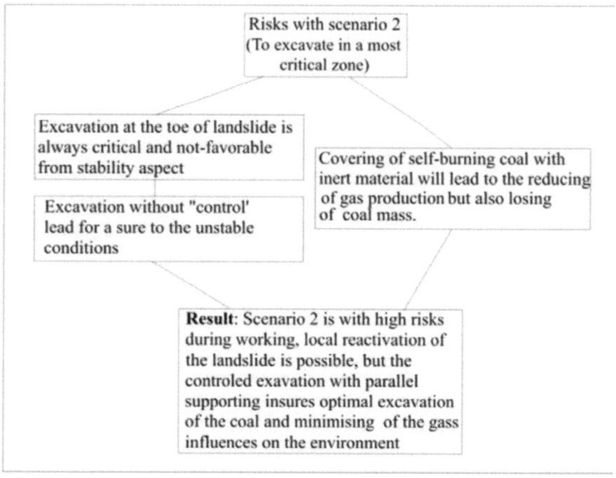

Figure 10. Presentation of main problems in Risk Scenario 1.

Both scenarios have possible negative influences on the environment and working conditions, but the main argument to accept Scenario 2 was the following.

- The process of self-burning leads to constant loosing of the coal mass and decreasing of the mechanical properties at the most critical toe zone from stability aspect.
- Covering of the zone of self-burning will lead to final closure of this zone.

The named arguments were a reason to apply an engineering solution, not a typical mining practice, and to accept Risk Scenario 2. Namely, it was decided that it is better to start with excavation, which will be analyzed in details with all possible negative consequences, as opposed to allow to lose a high quantity of coal in a process of self-burning, and finally to face the same situation—to have instability due to decreasing of the volume of the coal in the toe of the landslide. Shortly, the solution can be explained as a methodology of parallel excavation and supporting. In phases of decision-making, we used methodology of the so-called interaction-matrix method firstly introduced by [4]. The most important step in this methodology is to establish the objectives of the project and the analysis. The relevant problems are placed along the leading diagonal of conceptual interaction matrix. Then, all the interactions are established, and the problem structure is developed. An example of a relevant interaction scheme is in a form of Geotechnical Interaction Matrix and is presented in **Figure 11**.

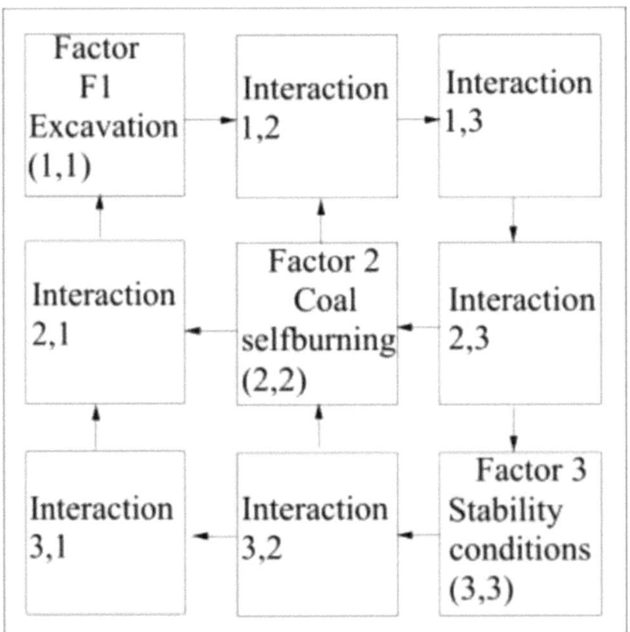

Figure 11. Conceptual matrix of interaction between tree basic factors.

F1 group of factors is related to the technology of excavation such as applied excavation and supporting method, depth of excavation, way of transportation, dewatering concept and so on. F2 group is related to the characteristics of self-burning process (area of burning, intensity, gas production, etc.). Group of factors F3 is related to the stability of the field, defined with movements of the masses and safety during the work. All possible interactions in a fist place are defined qualitatively, which is a very important step for engineering judgment and decisions. Explanations are given as follows.

- Interaction 1, 2 means that elements of the excavation can have an influence on the process of self-burning, because faster and efficient nearby excavation can stop the spreading of burning in wider areas.
- Interaction 1, 3 means that the elements of the excavation have a direct influence on the stability conditions, because correctly designed and applied technology of excavation create stable field conditions.
- Interaction 2, 3 means that the process of self-burning during longer time has an influence on the shear-strength parameters and leads to possible unstable conditions (beside other negative influences).
- Interaction 3, 2 shows that the stability of the field is the governing element, which affects possible access to zones of self-burning.
- Interaction 3, 1 means that stability of the field affects the way of excavation technology in numerous ways.
- Interaction 2, 1 means that the process of self-burning influences the excavation process, because of difficulties in access and in heavy working conditions.

It is obvious that such "simple" matrix shows several complex mutual influences between the environment and the engineering activities, and all of this shall be incorporated in design.

Based on this approach, detailed stability analyses were prepared for some representative profiles [6]. The software package SLIDE 5, product of RocScience, is used. The input parameters are defined earlier during the phases of investigations as well as with back analyses. The main properties are given in **Table 3**.

Material type	Cohesion C (KPa)	Angle of internal friction ϕ (o)	Unit weight γ (kN/m^3)
Disturbed trepel	0	13	15.64
Coal-like clay	0	9–10	16.63
Silty sands	0	21	21.25
Silts with high plasticity	0	11	19.5
Crushed coal in a sliding mass	15	25	11.61
Gneiss	200	50	26

Table 3. Main physical and mechanical parameters of the materials in a sliding mass.

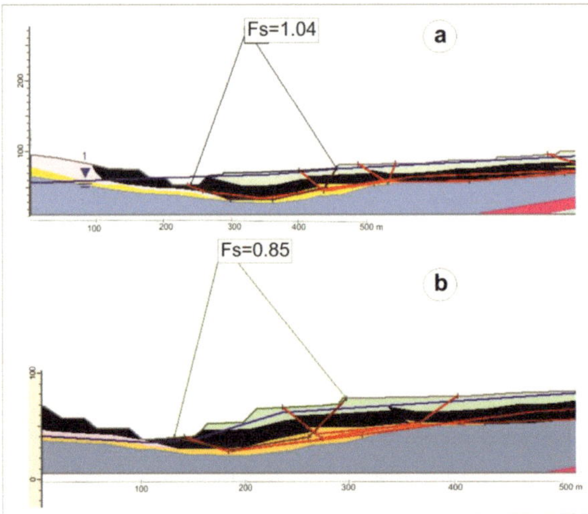

Figure 12. Typical outputs from stability analyses for initial phases of excavation; (a) before any kind of engineering activities; (b) hypothetical case without parallel support of the excavated zone.

Different phases of excavations and scenarios are involved in calculating. For example, in **Figure 12a**, we illustrate a value of safety factor (FS = 1.04) before any kind of engineering activities. In **Figure 12b**, we illustrate a hypothetical value of safety factor. This is a case, if we have a case without parallel support of the excavated zone when the safety factor is bellow FS = 1.

In practice, this case can be explained as a state of allowable deformations in a term of slow (controlled) sliding, which is expected during initial phases of excavations.

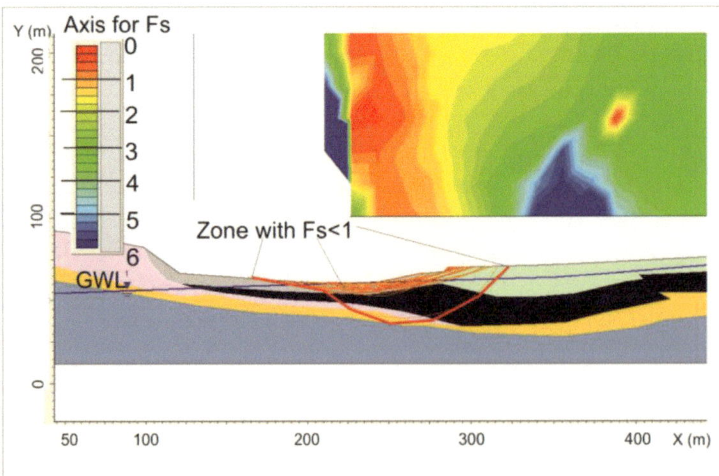

Figure 13. Simulation of stability if the process of self-burning was not stopped.

In **Figure 13**, we give an estimation of the long-term influence of the self-burning process. For the case in **Figure 13**, it was estimated that, in long-term, the upper zones of coal will be transformed into coal ash. Minor unit weight and internal friction angle are used in analyses. Results show that "new" sliding surfaces can be expected with values of safety factor smaller than 1 (unstable state).

Figure 14 explains cases of parallel excavation, support, and decreasing of artesian pressure, when the safety factor has values FS = 0.98 and FS > 1.1, respectively.

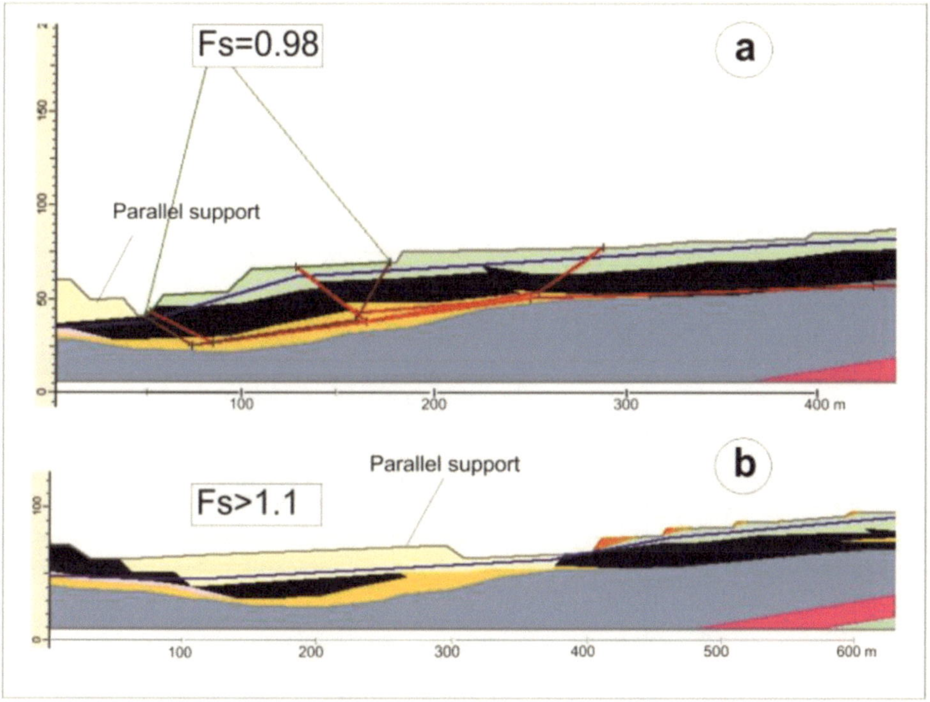

Figure 14. Analyzed cases with (a) parallel excavation and support; (b) with parallel excavation, support and decreasing of artesian pressure.

It can be concluded that for all variants, the values of safety factor that are usually not allowed in the mining practice. On the other side, the designers went into the calculated risk to excavate some quantity of the coal from one side and from the other to stop the process of self-burning. The main prerequisite to accept this risk was to apply all measures of surface dewatering and to have all time visual and geodetic observations during the work for control of possible rapid movements.

The excavation was conducted with discontinued type of equipment, which can be evacuated in a fast way if necessary. It can be noted that to date, in total, about 4,000,000 tons of coal are already excavated at this critical zone, with parallel support at the toe. As expected, minor gradual movements were observed during excavations.

5. Conclusions and recommendations

The given analyses are example, but in practice, it is sometimes necessary to deal with unusual cases that face high risks. This must be not a rule but only exceptions from rules.

All approaches in investigation and design shall completely be adapted to the characteristics of the natural environment; it is not possible to define the physical model of the terrain. Thus, we suggest the methodology of analysis presented in **Figure 15**.

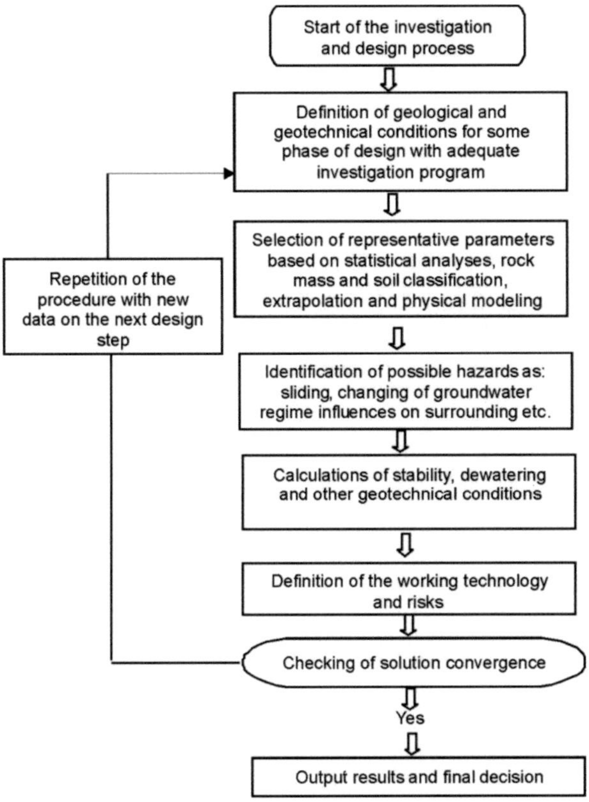

Figure 15. Suggested methodology for hazard and risk estimation at surface coal mines.

In every case, this article clearly shows that it is fundamental for successful design of each engineering activity to be acquainted in detail with the properties and conditions of the work and natural environment, possible hazards, and risk estimations.

The physical model of the terrain must be the basis for all numerical and mining analyses. We suggest using the interaction matrix method, as a useful approach in decision-making. Defined interactions are a good basis for complex analytical and numerical analyses, where the interactions can be defined with all necessary outputs (safety factors, stress-strain conditions, groundwater quantities, etc.).

Such approach can be adapted for numerous engineering problems, but it is necessary to have a team of specialists in mining, geological, and geotechnical engineering to solve such heavy engineering problems in an appropriate way.

Author details

Milorad Jovanovski* and Igor Peshevski

*Address all correspondence to: jovanovski@gf.ukim.edu.mk

Faculty of Civil Engineering, University Ss. Cyril and Methodius, Skopje, Republic of Macedonia

References

[1] Jovanovski M., Papic J., Dambov R., Pesevski I.: Stability aspects of excavation in landslide zone for a coal mine Suvodol, R. Macedonia, 3rd Balkan Mining Congress, Izmir, Turkey, 2009.

[2] Gapkovski, N., Jovanovski, M., Manasiev, J., Martinovic, A.: Report from the additional geotechnical investigations on the N.E. part of the coal mine "Suvodol" – Bitola, Faculty of Civil Engineering, Skopje, 1996.

[3] Jovanovski, M., Gapkovski, N.: Methodology and results from investigation of groundwater condition on the N.E. part of coal mine "Suvodol" – R. Macedonia, 6th International Mine water Association Congress, Bled, Slovenia, 1997.

[4] Hudson, J.A.: Comprehensive rock engineering, Pergamon Press, Oxford, England, 1993.

[5] Jovanovski, M., Donevska, K., Peshevski I.: Design of the dewatering at a coal mine Suvodol, Faculty of Civil Engineering, Skopje, 2007.

[6] Jovanovski, M., Papich, J., Peshevski I.: Design of the stability of the slopes at the coal mine "Suvodol", Faculty of Civil Engineering, Skopje, 2007.

[7] Jovanovski M., Popovska C., Donevska K., Peshevski I.: Interaction matrix method in hydrogeological analyses at coal mines, 11th International Symposium on Water Management and Hydraulic Engineering, Ohrid, R. Macedonia, September 04–09, 2009.

[8] Panovski A., Djorlev, S., Jovanovski M.: Design of an excavation in an landslide zone of the open coal mine Suvodol, RI Ruding, Skopje, 2007.

Chapter 3

Human-Induced Geo-Hazards in the Kingdom of Saudi Arabia: Distribution, Investigation, Causes and Impacts

Ahmed M. Youssef, Hasan M. Al-Harbi,
Yasser A. Zabramwi and Bosy A. El-Haddad

Additional information is available at the end of the chapter

http://dx.doi.org/10.5772/66306

Abstract

Different types of geological hazards are induced by human activities in the Kingdom of Saudi Arabia (KSA). These geological hazards include land subsidence and earth fissures, sinkholes, expansive soils, and flash floods. A wide variety of recent geological hazards have been reported in several areas, causing significant human and property losses. Human activities, most notably groundwater extraction, infrastructure development, and agricultural activities, have induced unstable conditions. This chapter provides an overview of the human-induced geological hazard in the KSA, mainly earth fissures and sinkhole, which represent a scarcely explored topic. This work identifies the main types of human-induced geological-hazard formations, distribution, causes, and impacts, illustrated through several case studies in the KSA.

Keywords: earth fissures, sinkholes, human induced, KSA

1. Introduction

Most frequent types of geological hazards observed in the Kingdom of Saudi Arabia (KSA) can be categorized into sand accumulations; land subsidence and earth fissures; flash floods; problematic soils; slope-stability problems; karst problems; faults; volcanic activities; and earthquake hazards. These hazards can be either natural or human-induced. Most of these hazards have been recorded in the KSA [1]. In this chapter, two types of geological hazards that are induced by human activities will be discussed. These hazards include earth fissures and sinkholes. Different types of problems associated with earth fissures and sinkholes, including water leakage in reservoirs, instability problems, flooding, infrastructure and urban-area damage, and loss of human life were recognized and evaluated by many authors [2–9]. The government of the KSA has also contributed to the extensive program of the agricultural activities for the past few decades [10]. These agriculture activities are considered as being of the largest water consumption, leading to the fast rate of depletion of nonrenewable groundwater. Analyzing the situation, it was found that:(1) during the period from 1976 to 1993, the area of agricultural land increased from 1600 to 32,000 km^2 and was irrigated with fossil groundwater; (2) water extracted from the aquifer was doubled between 1985 and 1990, reaching around 21,000 million cubic meters (MCM)/year. Consequently, a significant drop in the groundwater level was recorded in many regions (e.g., 100 m in the northwest area of the KSA); (3) it is expected that at the present rates of groundwater extraction, most of the fossil water will be depleted within 25–30 years. Earth fissures and sinkholes (related to carbonate and evaporite rocks) are considered to be the most-frequently observed and occurring geohazards in the KSA. The associated damage due to earth fissures and sinkholes are expected to increase in the future due to the anthropogenic alterations and the expansion of development. It was found that from years 2000–2010, the population has increased dramatically and a significant proportion of the population occupy karst areas of the eastern part of the country (e.g., Ar Riyadh) that leads to increase of vulnerability with respect to human-induced hazards. Different studies were performed in the KSA dealing with earth fissures and sinkholes [1, 11–18]. Many authors indicated that the main karst units in the Arabian Platform are Arab, Hith, Sulaiy, Aruma (Badanah, Zallum), Umm er Radhuma, Rus, Dammam, Dam, and Sirhan Formations [10, 12–14, 19].

2. Case-study group 1: earth fissures

2.1. Earth fissures' backgrounds

Earth fissures and the associated subsidence represent a major problem in different countries. Earth fissures and ground subsidence are related to the downward ground-surface movement compared with surrounding areas, ranging from strain cracks to large faults, starting from the ground surface in uncemented sediments [20, 21]. Earth fissures start from great depths below the surface, as a result of horizontal movement in the aquifers, because of excessive withdrawal (pumping) of the groundwater from the uncemented reservoir layers, due to loess soil, and earthquake activities [22, 23]. These earth fissures and subsidence could cause many problems in different urban and agricultural areas as well as damage infrastructure [24, 25]. Holzer [26] indicated that earth fissures can extend for a distance of tens of meters to kilometers due to tensile stresses. Under arid desert conditions, the shortage of groundwater resources and excessive pumping may cause continuous decline in groundwater levels [27]. When the aquifer is formed of unconsolidated sediments of high porosity and is interbedded with clay aquitards of low permeability and high compressibility, the rapid lowering of the groundwater level may also cause subsidence and possible ground failure in the form of earth fissures. Many authors documented that earth fissures and land subsidence can be related to groundwater withdrawal [28–35].

2.2. Human-induced earth fissures in the KSA

Earth fissures can be formed in loess soil due to water effect (rain, storms, floods, or leakage from agricultural irrigation channels and/or from neighboring houses). Earth-fissure and land-subsidence problems were reported in several areas in the Kingdom of Saudi Arabia causing damage to infrastructures, buildings, and agricultural areas (**Figure 1**). There are many examples of loess-related failures (earth fissures and subsidence) in the KSA such as the areas north of Jizan city, El-Darb Area, and North of Al-Nai village of Hail Region [1]. Many areas in the KSA are suffering from excessive groundwater extraction and are consequently subjected to land subsidence and earth fissures [1, 36–43]. Different types of earth fissures were detected in the KSA according to various reasons, among them are (a) earth fissures associated with groundwater extraction for agricultural development such as in Wadi Najran, Wadi El Dawather, Hail Region, Qasim Region, and Al Jouf Region; (b) earth fissures that are related to clay deposits (swelling and compressed clay deposits) such as in Hail, Al Qasim, and Al Jouf Regions;(c) earth fissures due to the Khabra deposits, which appear due to the drying effect such as in Al Jouf, Hail, and Al Qasim Regions; (d) earth fissures that are related to geological structures and groundwater withdrawal; and (e) earth fissures that are due to earthquake effects (El Shaqa area, northwest of El Madinah).

Figure 1. Earth fissures distributed in different areas in the KSA.

2.3. Earth fissures in Tabah village

2.3.1. *Tabah area*

The old village of Tabah is located ~70 km southeast of the Hail city at a latitude of 27° 02′N and a longitude of 42° 10′E (**Figure 2a**). It is laid on Harrat Hutaymah in an ancient volcanic crater (**Figure 2b**). This crater has dimensions of ~1.7 km by ~1.5 km and is filled with fine sediments and gravel (eroded from the surrounding volcanic tuffs), subsequently, filled with water forming a groundwater aquifer. The aquifer has been used since few decades until now by the village for drinking and irrigation. The crater has a low rim where weakly lithified tuff is exposed. Some Precambrian rocks and dikes are exposed inside the crater due to the erosion of tuff materials to fill the volcanic vent. The earth fissures are located in the old village of Tabah, which lies about 1 km southwest of the new Tabah village (**Figure 2b**). These earth fissures have been recognized and recorded since 1984. Extensive field investigations have shown the presence of earth fissures in different types, lengths, shapes, and directions (**Figure 3a**). Most of them are shown as ring shapes, forming a concentric zone along the margins of the volcanic vent (with a dimension of ~0.9 by ~0.7 km). They are deep, wide-open, and long fissures (more than 4 m deep, 3 m wide, and 600 m long). They spread in the floor of the village, agricultural areas, and cut through buildings (**Figure 3b**). Most nearby buildings and agricultural areas were damaged by these earth fissures. The presence of the earth fissures in the area leads to migration of most of the population to other areas (New Tabah village). There are some previous studies conducted in this area such as detailed in Refs. [24] and [36]. These studies indicated that the first earth fissures that were 120 m long were observed in 1981. Roobol et al. [24] mentioned that extensive earth fissures (as long as 500 m long) happened in 1984, causing damage to buildings. The present typical situation of these earth fissures is shown in (**Figure 2b**). The earth fissures in the area are shown as concentric rings (zone). Along this zone, most of the agricultural areas, village buildings, and concrete walls of the village cemetery were damaged. These earth fissures are associated with ground subsidence in some

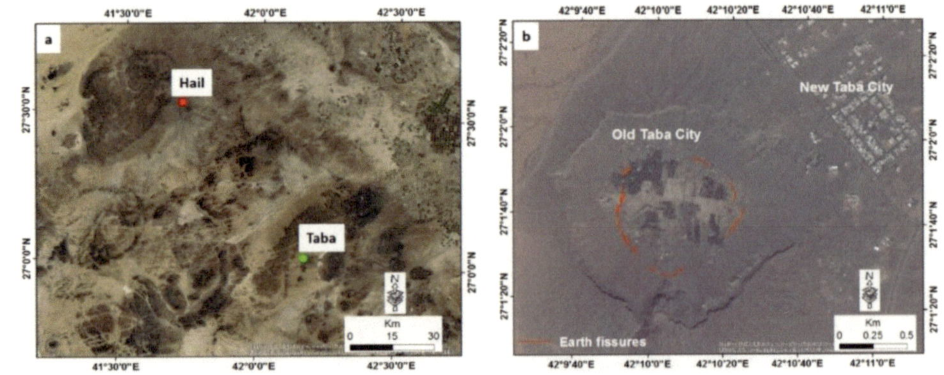

Figure 2. (a) Location of Tabah village related to the Hail city; (b) Location of earth fissures in the old Tabah village in relation to the New Tabah village.

http://dx.doi.org/10.5772/66306

parts, showing a vertical displacement (up to 1.5 m). We believe that these earth fissures and ground subsidence are not yet completed and will continue in the future.

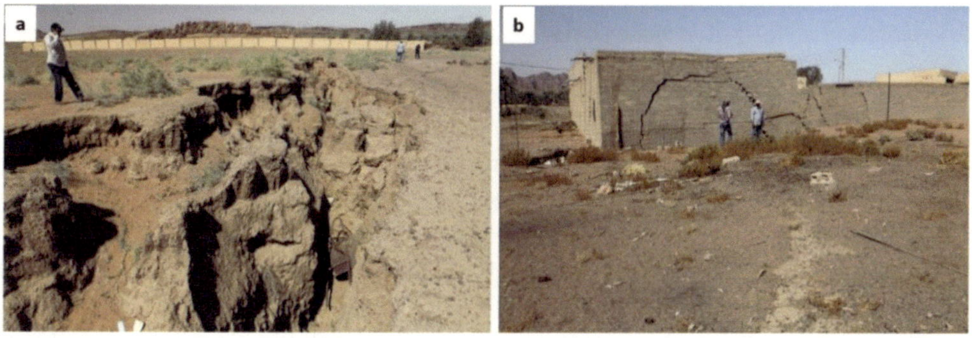

Figure 3. Photographic pictures showing (a) Earth fissures' distribution, (b) Impact of earth fissures on buildings.

2.3.2. *Topography and geology*

Analysis of satellite image and topographic map of the study area indicated that the earth fissures are located between elevations of 1020 m and 1030 m from the mean sea level. Geologically, the study area consists of the following geological units from the youngest to the oldest as follows (**Figure 4**): (1) Khabra deposits (Qk), including silt, clay, and some sand; (2) Reworked volcanic ash deposit (Qa) consisting of a redeposition of ancient volcanic deposits;

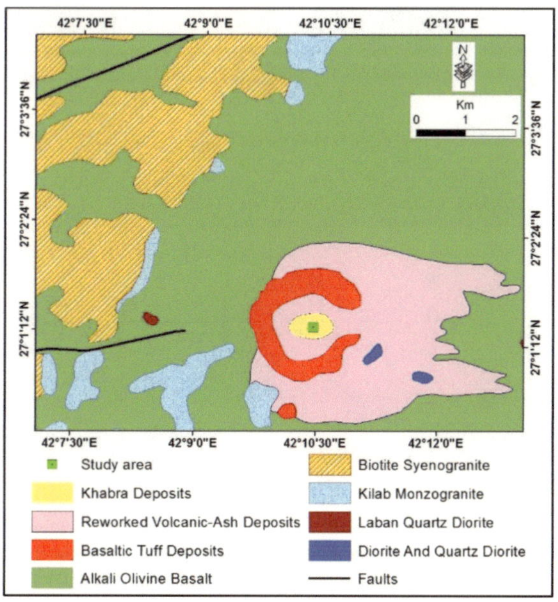

Figure 4. Surface geological map of the study area and its surrounding.

(3) Basaltic tuff deposit (Qt); (4) Alkali olivine basaltic rocks (Qb); (5) Biotite syenogranite rocks (sgb); (6) Kilab mozogranite rocks (kmg); (7) Laban quartz diorite rocks (lqd); and (8) Diorite and quartz diorite rocks (di).

2.3.3. Geophysical investigation and analysis

Geophysical investigation using the electrical-resistivity method was performed and is presented in this chapter. Electrical-resistivity tomography (ERT) represents commonly used geophysical techniques for the detection of earth fissures, which has been widely used in the KSA, producing satisfactory results. ERT is particularly useful in areas with significant resistivity contrasts. For measurements, four electrical-resistivity lines were performed in the old Tabah village (**Figure 5**). The first electrical line has a west-to-east direction with a total length of 285 m and spacing of 1 m and 3 m between electrodes, while the remaining three lines have south-to-north direction and lengths of 960 m, 470 m, and 470 m for the second, third, and fourth lines, respectively. The spacing between the electrical poles is 10 m each.

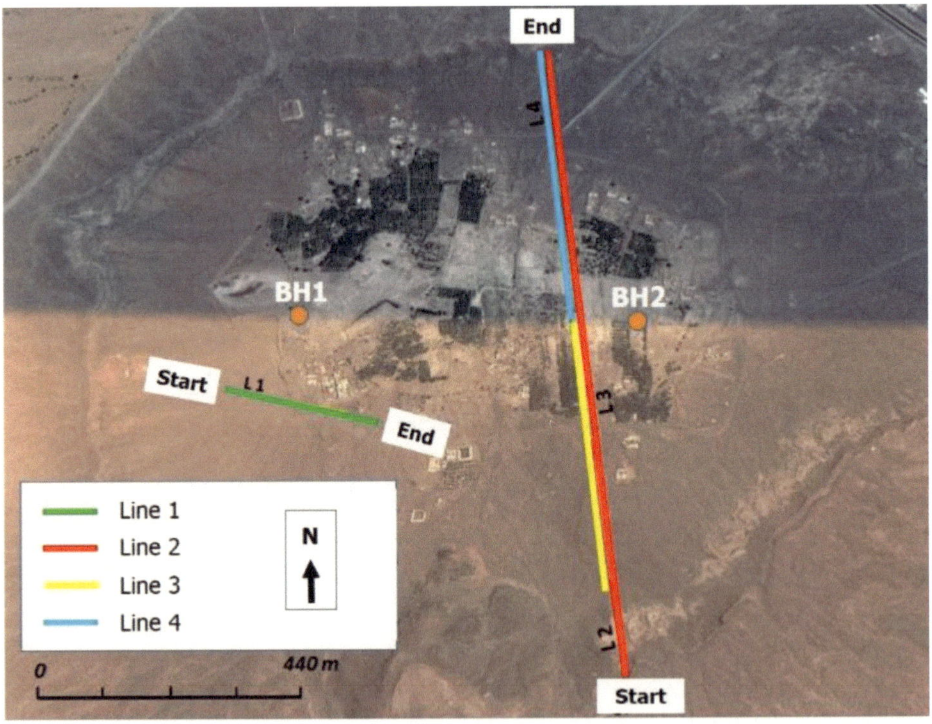

Figure 5. Distribution of geophysical lines and boreholes in the study area.

Line 1 (L1) moves across the earth fissures with a west-to-east direction. It applied with 1 m electrode spacing (**Figure 6a**) and 3 m spacing (**Figure 6b**). Information collected using this line shows that the soil in the region, which consists of Quaternary and volcanic ash sediments, has low electrical resistance up to the maximum depth of the profile (40 m). The decrease in the electrical resistance values is related to the water-saturation effect. It reveals the presence of vertical fissures in the middle part of the electric line (L1) (**Figures 6a**, **b**), which represents the contact line between the water-saturated soil and the Precambrian complex rocks, which has high electrical resistance. These hard rocks begin from 5 m under the surface and extending up to 40 m deep.

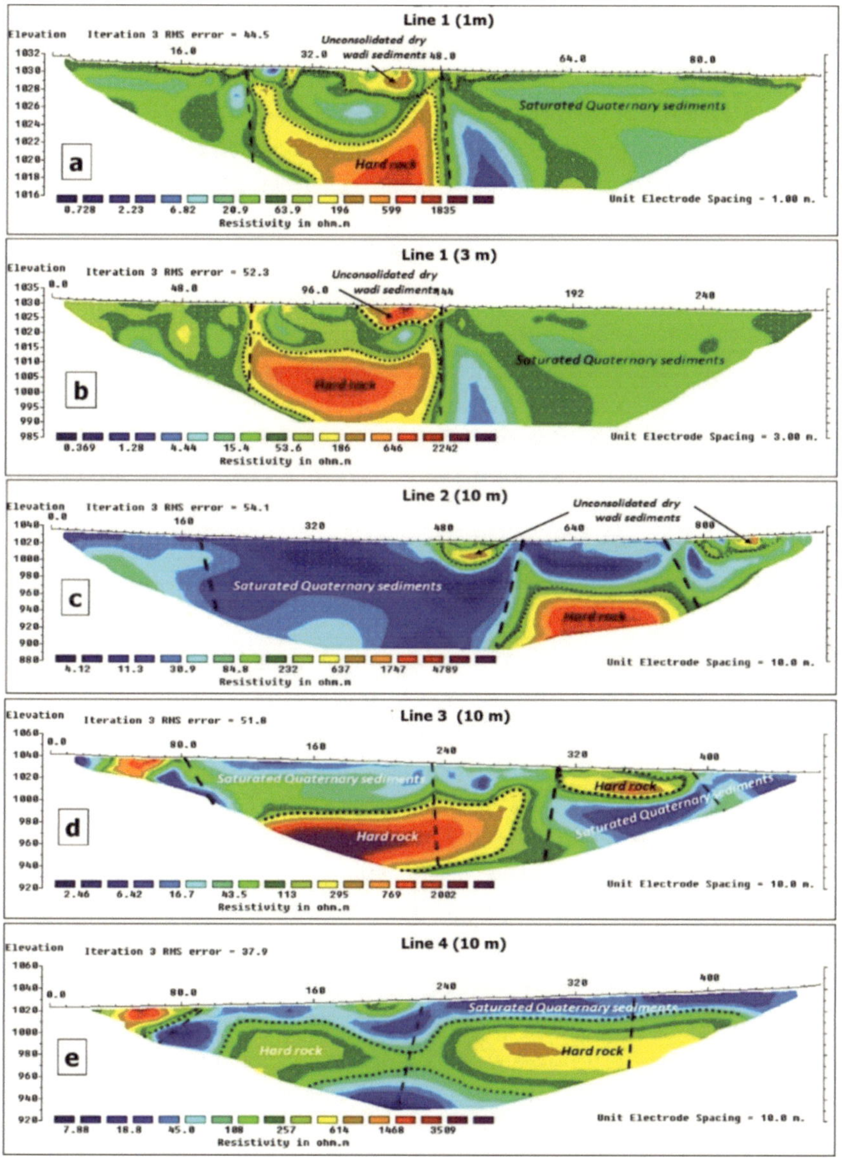

Figure 6. Electrical-resistivity profiles of different lines:(a) Line 1 with 1 m electrode spacing; (b) Line 1 with 3 m electrode spacing; (c) Line 2 with 10 m electrode spacing; (d) Line 3 with 10 m electrode spacing; and (e) Line 4 with 10 m electrode spacing. Note: horizontal axes do not have the same scale.

Line 2 (L2) moves across the earth fissures with a south-to-north direction with a total length of 950 m, 10 m electrode spacing, and reaches up to 130 m depth (**Figure 6c**). It is clear that there is a large underground reservoir with a large thickness as the electrical resistance values are low. The end of the profile shows the presence of high-resistance materials (solid rocks) at a depth of 50 m, extending beyond the total depth of the profile (130 m deep). Different deep earth fissures were detected along the profile and appeared at the contact between sediments and hard rocks (**Figure 6c**).

Line 3 (L3) moves across the earth fissures with a south-to-north direction with a total length of 470 m, and 10 m electrode spacing (**Figure 6d**). It was found that this profile reflects the same phenomena that have been monitored in Lines 1 and 2 where there is a large zone of water-saturated sediments (with low electric resistance), which intruded withy hard rocks (with high electric resistance). The hard rocks appear at a depth between 30 m and 50 m in the middle of the profile. Various deep earth fissures were detected along the profile at distances of 80 m, 230 m, 310 m, and 390 m from the start point of the line (**Figure 6d**). These earth fissures mostly appear along the contact between soil and rocks.

Line 4 (L4) moves across the earth fissures with a south-to-north direction with a total length of 470 m and 10 m electrode spacing (**Figure 6e**). It was found that this profile reflects the same phenomena that have been monitored in Lines 1, 2, and 3 where there is a large zone of water-saturated sediments (with a low electric resistance), which intruded withy hard rocks (with high electric resistance). Hard rocks appear as a horizontal layer at a depth of 10 m from the surface. Various deep earth fissures were detected along this profile (L4) at distances of 100 m, 210 m, and 345 m from the start point (**Figure 6e**). These earth fissures mostly appear along the contact between soil and rocks.

2.3.4. Geotechnical investigation and analysis

Two boreholes were drilled in the study area to investigate the subsurface soils and rocks (**Figure 5**). Analysis of the boreholes indicated that the sediment layer is characterized by alternation of sand with silt, clay with silt, sand with clay, and silt and some gravels and different rocks (**Figure 7**). These deposits have color ranging from light brown to brown and a thickness of 19 m (BH 1) to 144 m (BH 2). According to the unified soil classification system (USCS), these sediments include (a) clay with silt (CL–ML), which are characterized by liquidity-limit values ranging 18–28, plastic-limit values ranging 12–21, and plastic-index values ranging 5–7; (b) sand with silt (SM) and sand with silt and clay (SC–SM), which are characterized by a liquidity-limit value of 13, plastic-limit value of 9, and a plastic index value of 4.

Rocks are characterized by the following: (a) clay stone is characterized by gray to white color, includes some clay and silt, highly fractured, the rock-quality designation (RQD) ranging from very poor to poor (0–40) percent of the recovery ranging from 40 to 100%, and the uniaxial compressive strength ranging from 0 MPa to 4.3 MPa; (b) sandstone is characterized by brown color, highly fractured, the rock-quality designation (RQD) ranging from very poor to fair (20–66), the recovery percent of 100%, and the uniaxial compressive strength ranging from 4.6 MPa to 7.4 MPa; (c) siltstone is characterized by brown color, moderately to highly fractured, the rock-quality designation (RQD) ranging from very poor to fair (0–40), the recovery percent ranging from 90 to 100%, and the uniaxial compressive strength ranges from 1.4 MPa to 7.2 MPa; (d) igneous rock is characterized by brown color, fine grained, slightly to highly fractured, the rock-quality designation (RQD) ranging from very poor to fair (0–74), the recovery percent of 100%, and the uniaxial compressive strength ranging from 1.1 MPa to 15.3 MPa.

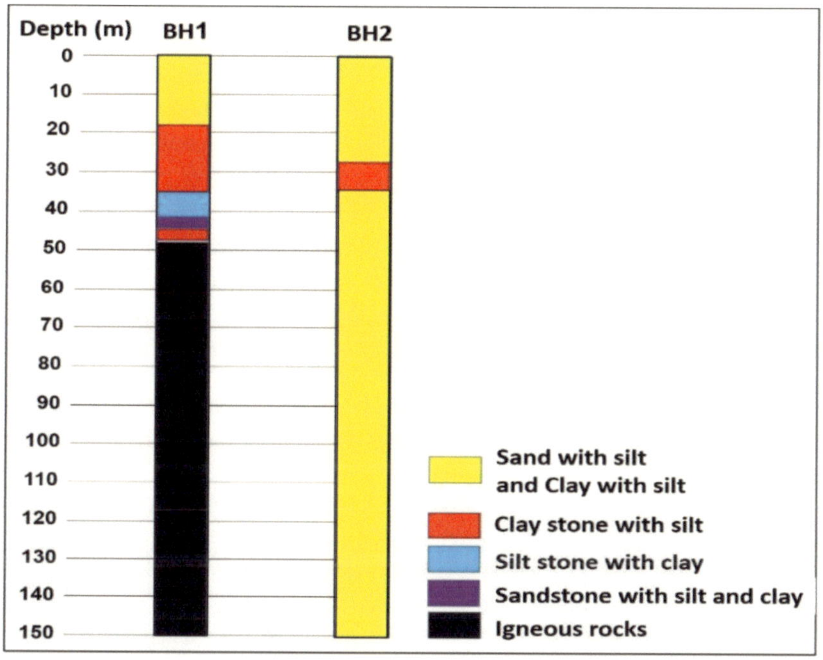

Figure 7. Lithology of boreholes.

2.3.5. *Causes of earth fissures*

Throughout the abovementioned field, remote-sensing maps, geological, geophysical, and geotechnical investigations and analysis, the earth fissures and ground subsidence in Tabah area have been observed to be occurring since many decades ago and still continuing. The study indicated that these earth fissures were located inside the volcanic crater, which was deep and open. The volcanic tuffs and agglomerate surrounding the vent were eroded due to the effect of rainfall and filled the vent. These earth fissures are related to the development of agricultural activities, which mainly depend on groundwater withdrawal. Most causes of earth fissures can be discussed as follows:

a. The area is characterized by the presence of agricultural areas since few decades ago, which mainly depend on the groundwater aquifer in the area (**Figure 8**). According to the information obtained from residents of the area, the groundwater level was near the earth's surface (<50 m below the surface) 30 years ago. As a result of drilling of a large number of groundwater wells (for domestic use, irrigation, and road building, which in turn led to extensive withdrawal of huge amounts of groundwater), the water level declined to the level of 120 m in 1985 and now lower than 160 m below the earth's surface.

b. Subsurface materials in the Tabah area is characterized by breccia, gravel, sand, silt, and clay as well as different rocks such as claystone, siltstone, and sandstone. The sediments extend to a depth greater than 150 m at Borehole 2 and may increase in the middle of the area. The thickness of these sediments decreased outwards to about 48 m in Borehole 1 and decreased outwards until igneous rocks appear on the surface.

c. As a result of the topography irregularity, the thickness of the sediments increased toward the centre of the area and decreased outwards where the solid rocks become near the surface especially on the sides of the area. As a result of the groundwater withdrawal, the water level significantly declined, leading to the compression of the sediments. According to the differential settlement of these sediments, along the edges, the amount of settlement is small and increases toward the centre of the area. This leads to the appearance of earth fissures at the contact surface between hard rocks and deep sediments. Some of these fissures have vertical displacement. **Figure 9** shows a model for the earth-fissure development. The geophysical and geotechnical investigations confirmed this theory as the cause of earth fissures.

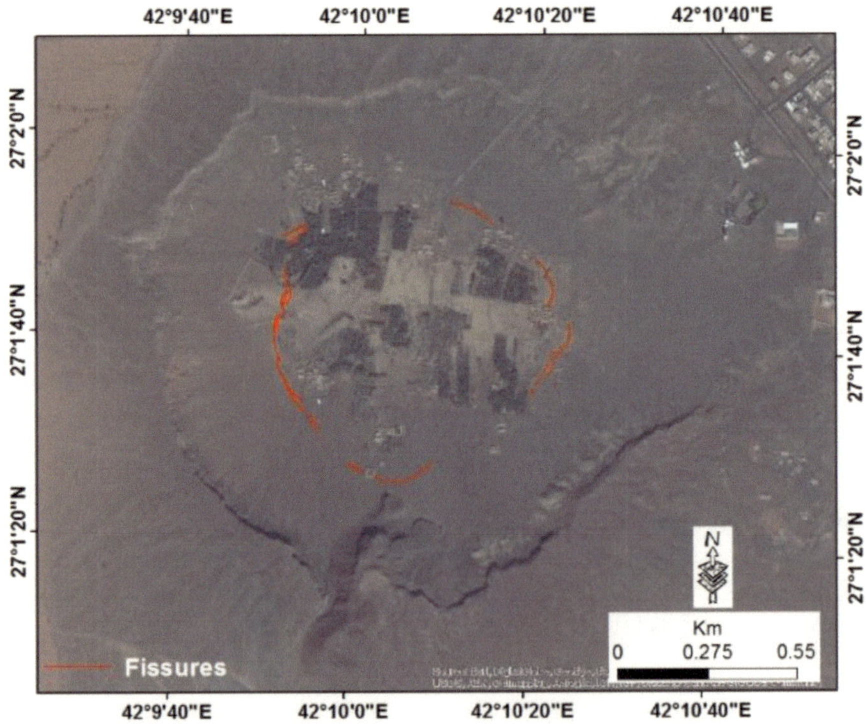

Figure 8. An example of agricultural activities in Tabah area.

http://dx.doi.org/10.5772/66306

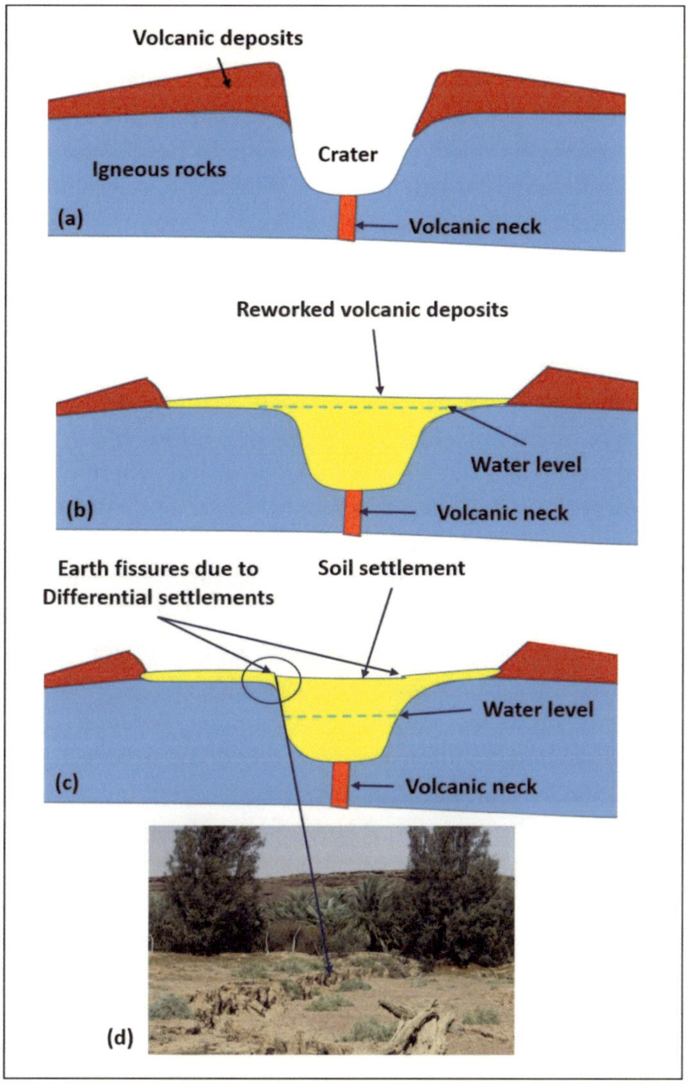

Figure 9. Earth fissure development model in Tabah area (modified after Roobol et al. [24]: (a) Before erosion of volcanic deposits; (b) After erosion of volcanic deposits and refill, the vent, water table is very shallow; (c) Current condition, water table ~160 m deep and earth fissures occurred; and (d) Photograph showing current earth fissures.

3. Case-study group 2: sinkholes induced by human activities in the KSA

3.1. Karst backgrounds

Karstic rocks cover a large area of the Kingdom of Saudi Arabia, mainly in the eastern and northern parts [44]. Sinkholes, the main manifestation of Karstification on ground surface, represent one of the natural geohazards. Most damaging incidents related to the presence of sinkholes are induced by human activities. Many cases in the KSA illustrate that there is increasing human impact on the natural environment and aquifers due to the rise in human development and activity. Interstratal dissolution of these formations and the subsidence of the overlying sediments has generated numerous large subsidence depressions and sinkholes [16, 45, 46]. In fact, Ar Riyadh, the capital of the Kingdom, is located on a large inactive subsidence depression related to interstratal evaporite dissolution [46–48]. A similar situation is found in other sectors of the country underlain by the Late Cretaceous Badanah and Zallum formations with interbedded evaporites at shallow depth [49]. Here, extensive tracts are riddled by khabras that correspond to subsidence depressions generated by interstratal karstification of evaporites and the subsidence of overlying sediments [50]. Youssef et al. [10] indicated that a large number of new sinkholes have been discovered in recent years, notably in Ar Riyadh area, the Al Summan Plateau northeast of Ar Riyadh, in an extensive belt south of Ar Riyadh extending as far as Sulayyil, in the eastern and northern provinces, and in Jazan area on the western coast. They indicated that most of the recently documented sinkholes are related to human activities that may cause dissolution and/or subsidence processes (groundwater withdrawal, irrigation, water leakage, and overloading), suggesting a significant induced component. Various types of sinkholes were recorded in the KSA that are related to human activities [10]. These sinkholes were categorized based on the classification presented by Gutiérrez et al. [51–53]. Gutiérrez et al. [51–53] classified sinkholes into two types: solution sinkholes and subsidence sinkholes. They indicated that (1) solution sinkholes are shallow, enclosed depressions generated by differential lowering of the surface in karstic rocks; and (2) subsidence sinkholes are resulting from subsurface dissolution and downward gravitational movement of the materials. The subsidence sinkholes in the adapted classification use two terms: one related to material affected by subsidence (cover, bedrock, and caprock), and the second term represents the subsidence mechanism (collapse, suffosion, and sagging). In the KSA, different types of subsidence sinkholes were recorded including cover- and caprock-collapse sinkholes, cover-suffosion sinkholes, sagging sinkholes, and complex sinkholes. Cover refers to unconsolidated deposits; caprock refers to nonkarstic rocks; collapse indicates the brittle deformation of soil or rock material; suffosion is the downward migration of unconsolidated cover deposits through conduits; and sagging is the downward bending of ductile sediments. Complex sinkholes, on the other hand, involve more than one material type and different subsidence mechanisms. In the current section, different examples of these sinkholes in the KSA will be discussed in detailed:

1. Cover-collapse sinkholes case study, Al Jouf farm sinkhole: this type of sinkhole is a cover collapse sinkhole, located at a latitude of 29°46′43.68″N. and longitude of 38°27′37.02″E. This sinkhole has 40 m diameter and 15 m depth. It occurred in 2006 within a farm circle (**Figure 10a**). Geologically, the limestone bedrock of the Sirhan formation is overlain by a thick Quaternary cover (Aeolian sands capped by a quartz-rich gravel sediments). This sinkhole is formed due to excessive groundwater extraction, which started in 1989. Recent measurements of the water table indicated that the water level is at about 205 m deep from the ground surface. The drawdown of the water table has reached ~100 m.

2. Caprock-collapse sinkhole case study, Al Issawiah sinkholes: four sinkholes were recognized in Al Issawiah area, and some of them are caprock-collapse sinkholes. One of them is located at a latitude of 30°43′30.5″N. and longitude of 38°06′01″E. It is a subcircular collapse sinkhole with a diameter of 27 m and a depth of 23 m. It is opened on a basaltic caprock (**Figure 10b**). Geologically, the Sirhan formation in this area consists of friable calcareous sandstone, limestone, and shale, unconformably overlain in some areas by basalt of the Harrat Al Harrah lava field [54]. The basalt is covered by a thin silty soil plus discontinuous sand dunes and residual basalt boulders. In other areas, the basaltic layer disappears (**Figure 10c**). Field investigations indicated that the water table was exposed at the bottom of the sinkhole; with increasing groundwater exploitation because of increasing irrigation crops, the depression became dry and the water table declined.

3. Cover-suffosion sinkholes: cover-suffosion sinkholes develop in areas with karstified bedrock covered by an unconsolidated soil. These cover deposits may migrate downward through dissolutional conduits and enlarged joints. This leads to the progressive settlement of the ground surface. This type of sinkhole was recorded in Al Khafji area with a diameter of 70 m (**Figure 10d**). Geologically, the area consists of limestone-bearing bedrock of the Hadrukh formation overlain by a thick low-cohesion sand-gravel cover. This type is characterized by an ellipsoidal depression, 520 m long and 310 m wide. This sinkhole was potentially triggered because of lowering of the water table due to groundwater pumping from the Hadrukh formation and the underlying Dammam karst aquifer.

4. Sagging sinkholes: sagging sinkholes involve the progressive passive bending of sediments related to dissolution of underlying soluble material. This type of sinkholes appears in many areas in the KSA, which are underlain by various formations such as Jilh, Arab, Zallum, Badanah, and Umm er Radhuma formations. Most of these areas are characterized by khabras deposits, which are filled by Quaternary fine-grained deposits. Many authors indicated that these khabras correspond to large sagging sinkholes relating to differential, interstratal karstification of the gypsum beds and the progressive ductile bending of the overlying rock strata [47, 49, 50, 55]. Another example of sagging sinkholes was documented in the sabkha environment in Jizan area where there was subsurface dissolution of evaporites, frequently induced by artificial water input into the ground and caused ground settlement (**Figure 10e**) [1].

5. Complex sinkholes: these types of sinkholes result from the combination of two subsidence mechanisms (sagging and collapse processes). Many examples were documented in different areas of the KSA, including:

a. Aba Alwrood sinkhole located in Al Qasim Region at the latitude of 26°25′52.16″N. and longitude of 44°03′10.69″ E. This sinkhole is a sagging and collapsing sinkhole which is 10.5 m long and 7.5 m wide that occurred in 2010 (**Figure 10f**). Geologically, the gypsum bedrock is of Jilh Formation of Triassic-age [56].

b. Turaif sinkhole is located in Al Qasim Region, at the latitude of 26°49′26.65″N. and longitude of 44°09′46.83″E. Geologically, this sinkhole area consists of thinly bedded gypsum, limestone, and shale of the Triassic Jilh Formation [56]. The layers in the sinkhole display downward bending, indicating two mechanisms of subsidence, including sagging and collapsing (**Figure 10g**).

c. Bsita sinkhole is located in Al Jouf Region, at the latitude of 30°11′16.8″N. and longitude of 37°51′1.88″E. This sinkhole has an oval shape with a length of 100 m and width of 60 m (**Figure 10h**). Geologically, Palaeocene- and Eocene-age rocks of Mira formation are exposed in the area. These rocks consist of a thinly bedded silicified limestone and banded chert [54]. The studied evidence of this sinkhole indicates that the depression corresponds to a sagging-collapse sinkhole [10].

Figure 10. (a) Cover-collapse sinkhole in Al Jouf farm company area; (b) Caprock-collapse sinkhole in Al Issawiah area; (c) Caprock-collapse sinkhole in Al Issawiah area; (d) cover-suffosion sinkhole in Al Khafji area; (e) sagging sinkhole in Jizan area; (f) Bedrock-collapse and sagging sinkhole in Aba Alwrood area, Al Qasim region; (g) Bedrock-collapse and sagging sinkhole in Turaif area, Al Qasim Region; and (h) Bedrock sagging and collapse sinkhole at Bsita, Al Jouf Region.

3.2. Methods of sinkhole investigations

3.2.1. Using geologic and topographic maps

The Kingdom of Saudi Arabia is located on the southern part of the Arabian Plate. According to the geological maps of 1:250,000 scale, which covers the KSA, an important information on the distribution of different rock units can be obtained as shown in (**Figure 11**). The KSA consists of three main geological zones:

Zone 1: The Arabian Shield, situated on the western part of the KSA. It consists of Precambrian rocks, locally covered by Cenozoic lava flows [57].

Zone 2: The Arabian Platform, situated on the eastern and northern part of the KSA. It is characterized by a Phanerozoic sedimentary succession [58]. It includes a significant proportion of carbonate and evaporite karst formations.

Zone 3: The Quaternary sediments cover a narrow coastal strip along the margin of the Red Sea and the east margin of the KSA. It also covers some areas of Zone 2. According to this map, it was found that the carbonate and evaporate rocks cover a significant portion of the KSA, which is exposed on the surface and sometimes covered by other rock formations. Different types of sinkholes were documented and mapped along Zone 2 [41–43, 46, 49, 59–62].

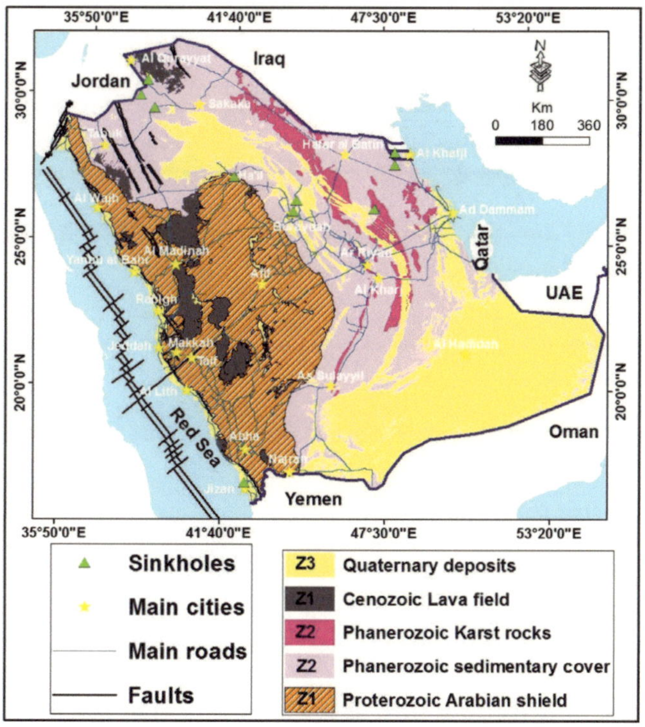

Figure 11. Geological map of the KSA showing areas of earth fissures and sinkholes.

According to topographic maps of 1:50,000 scale, different types of depressions related to dissolutions were detected and mapped. These depressions have local names (Dahls). The presence of these depressions can give a good indication of the presence of sinkholes and dissolution-induced subsidence depressions. Many studies were conducted on these depressions that are mapped in the topographic maps, and they verified that these depressions are related to dissolution and collapse sinkholes [10]. Some of these depressions (Dahls) on a topographic map of An Nu'ayriyah area is shown in **Figure 12**.

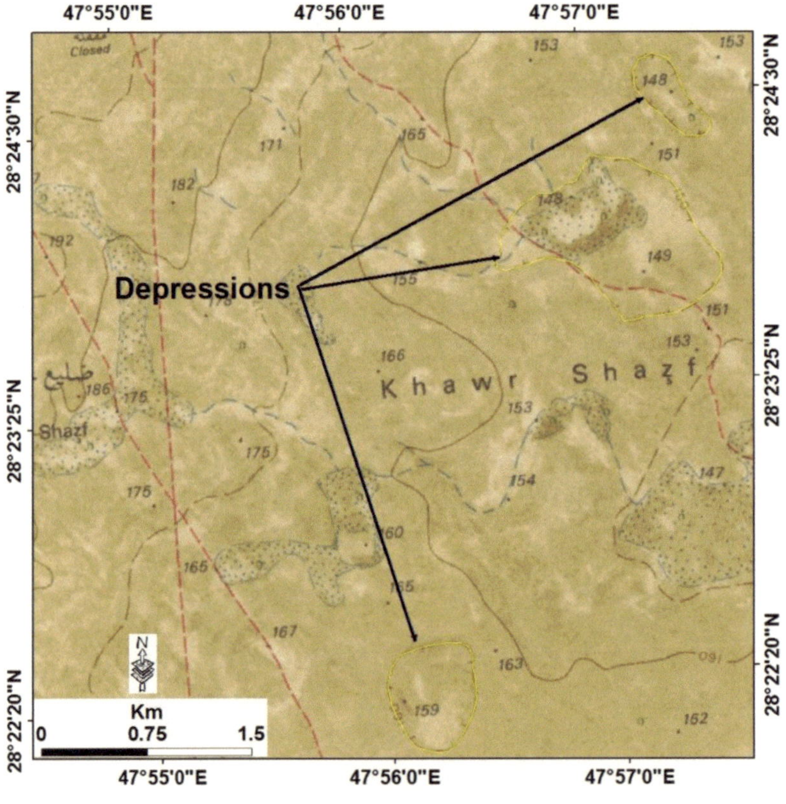

Figure 12. Topographic map of a small part of An Nu'ayriyah area showing dahls (potentially sinkhole areas).

3.2.2. Using remote-sensing images

One of the popular, nondestructive methods for detecting karst-related features and sinkholes is the use of remote-sensing data. These remote-sensing data can be used for recognizing different surface features that are related to karst phenomena using visual inspection. Different types of remote-sensing data have previously been used for distribution and recognition of sinkholes and karst-surface features such as aerial photographs [63, 64] and satellite images [10, 17]. In the KSA, aerial photographs are very rarely and in a limited manner used for detecting karst related features. However, Landsat images of 15 m resolution can easily be

http://dx.doi.org/10.5772/66306

prepared using the fusing technique to sharpen the resolution of 30 m bands 1, 2, 3, 4, 5, and 7 to enhance the resolution to 15 m resolution using panchromatic band 8. These data were used before to detect surface features (circular features, depressions, and ring structures) that are related to karstification at An Nu'ayriyah area (**Figures 13a, b**). Another set of remote-sensing data was used to detect sinkholes, including the imagery of professional Google Earth. These high-resolution images were used in different projects in the KSA and give valuable information about the presence of sinkholes and their time of occurrence. **Figures 13c, d** shows some examples of Google Earth high-resolution images with different sinkholes at Al Issawiah area.

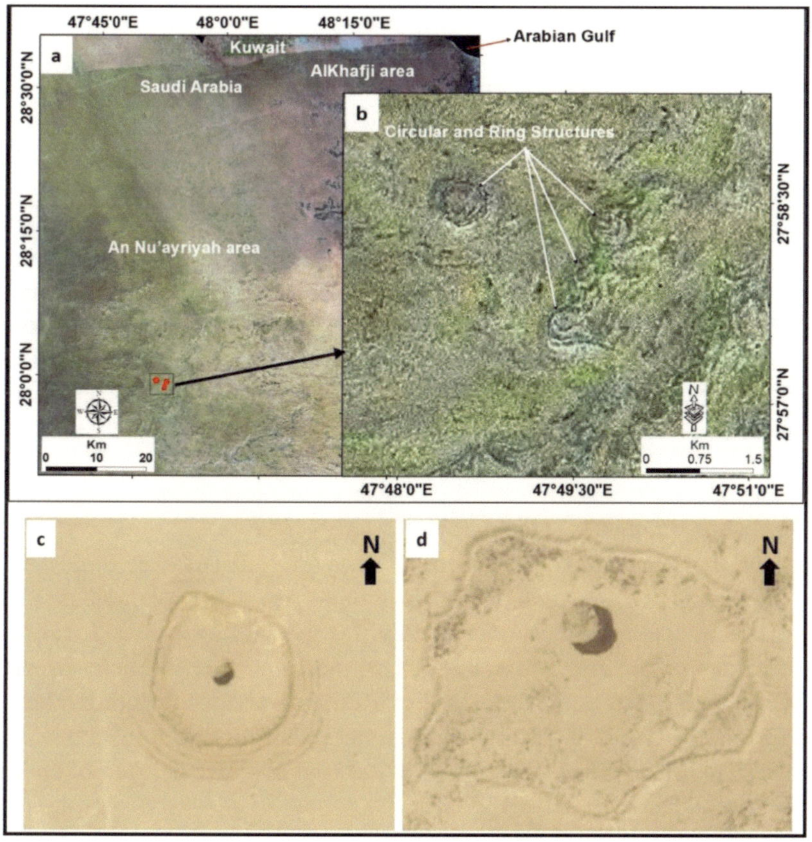

Figure 13. (a) Circular and ring structure features at An Nu'ayriyah area;(b) Circular and ring structure features at An Nu'ayriyah area; (c) Google-earth map of sinkholes at Al Issawiah area; (d) Google-earth map of sinkholes at Al Issawiah area.

3.2.3. Field investigation using trenching technique and borehole technique

The trenching technique has previously been applied in different investigations by many authors [10, 63, 65–68]. The trenching method depends on the excavation of trenches at the study site, mapping the exposed stratigraphy and structure as well as reconstructing the deformation history. The trenching technique has been applied in different fields such as ground instability (landslides) and sinkholes. Many examples were performed in the KSA to investigate the cover-collapse sinkholes. The trench has to be excavated perpendicularly to the edge of the sinkhole. Youssef et al. [10] mentioned that the walls of the trench must be cleaned and, then, logged on graph papers at a rescannable scale using an orthogonal grid of strings with a spacing of 1 m. Trenching method has various aims, including: (a) internal structure and subsidence mechanisms investigation (deformation style); (b) checking whether the sinkhole was an incipient-collapse feature or the reactivation of a pre-existing buried collapse depression; and (c) reconstructing the evolution of this presumably human-induced sinkhole to build a prognostic basis.

3.2.4. Detection of subsurface-karst features using geophysical techniques

Electrical-resistivity tomography technique is one of the most commonly used geophysical methods for the detection of cavities and buried sinkholes in karstic regions. This method depends on imaging the subsurface materials (according to the bulk electrical resistivity of each material type) by multielectrode systems [69]. Electrical-resistivity tomography (ERT) method gives excellent results in areas with significant electrical-resistivity contrasts. This is expected to occur in karstic terrains where most cavities, with considerable sizes, are filled with low electrical-resistivity deposits [10, 17, 18, 68, 70–74]. The electrical-resistivity tomography (ERT) is very common in the KSA, and reasonable results could be obtained using this method [11, 17, 18]. Electrical-resistivity and seismic techniques have been applied in different areas in the KSA and have been successful detecting sinkholes [10, 17, 75].

3.3. Causes of sinkholes

All sinkholes in the KSA are formed according to the presence of underlain cavernous limestone and evaporites. Formation of these sinkholes could be explained as subsidence and collapse processes that occurred above old cavities, probably formed during past pluvial phases. Gutiérrez et al. [53] and Youssef et al. [10] indicated that sinkhole-formation mechanisms can easily be initiated and accelerated by human activities (groundwater withdrawal, irrigation, and overloading). According to the available information of the sinkholes in the KSA, there are different anthropogenic factors that trigger the formation of sinkholes development including:

1. Excessive groundwater pumping from limestone-formation aquifers: this leads to rapid water-table decline and loss of buoyant support in the roof of pre-existing cavities and an increase in the effective stress. In addition, internal erosion processes could happen due to decline of water table. Different types of sinkholes have been recently recorded under this type of scenario, especially in the northern and central areas of the KSA (Al Jouf and Al Qasim Regions).

2. Dissolution of salt rocks and deposits due to infiltration of freshwater (leaking pipes) into the subsurface salt rocks and deposits: this process has been documented in Jazan area

where the old Jazan city was built on a salt dome and on the adjacent sabkha areas. The entire old Jazan city has been abandoned due to the severe damage by subsidence [1]. Irrigation and rainwater infiltration: this leads to an increase in the top-soil unit weight and a decrease in its strength and cohesion (silt and clay materials). That leads to a migration of cover deposits though underlying cavities.

3. Loading and man-made vibrations (static and dynamic loads): these lead to the collapse of unstable cavities. Many examples were documented in the KSA such as the sinkhole at Al Khobar city, in the eastern province, induced by the load of a vehicle on a road.

4. Excavation: This leads to the surface appearance of cavities underneath. One example of the appearance of underground cavities was reported during the construction of the wastewater treatment system in Ar Riyadh city where there are Arab, Hith, and Sulaiy formations (limestone and anhydrite).

Author details

Ahmed M. Youssef[1,2*], Hasan M. Al-Harbi[1], Yasser A. Zabramwi[1] and Bosy A. El-Haddad[2]

*Address all correspondence to: amyoussef70@yahoo.com

1 Geological Hazards Department, Applied Geology Sector, Saudi Geological Survey, Jeddah, Kingdom of Saudi Arabia

2 Geology Department, Faculty of Science, Sohag University, Egypt

References

[1] Youssef AM. Pradhan B. Sabtan AA. El-Harbi HM. (2011) Coupling of remote sensing data aided with field investigations for geological hazards assessment in Jazan area, Kingdom of Saudi Arabia, Environmental Earth Sciences, 65(1):119–130. Doi: 10.1007/s12665-011-1071-3.

[2] Crawford NC. (1984) Sinkhole flooding associated with urban development upon karst terrain: Bowling Green, Kentucky. In Proceedings of the First Multidisciplinary Conference on Sinkholes, ed. Beck BF. Balkema AA. Rotterdam. pp. 283–292.

[3] Gao Y. Luo W. Jiang X. Lei M. Dai J. (2013) Investigations of large scale sinkhole collapses, Laibin, Guangxi, China. In: LandL. Doctor, DH, Stephenson. JB (Eds). Sinkholes and the Engineering and Environmental Impacts of Karst. National Cave and Karst Research Institute. pp. 327–331. Carlsbad, NM: The National Cave and Karst Research Institute (NCKRI). www.karstportal.org/sites/karstportal.org/files/KIP-00011836.pdf

[4] Gutiérrez F. (2010) Hazards associated with karst. In: Alcántara I and Goudie A. (Eds.). Geomorphological Hazards and Disaster Prevention. Cambridge University Press. Cambridge. pp. 161–175.

[5] Gutiérrez F. Mozafari M. Carbonel D. Gómez R. Raeisi E. (2014a) Leakage problems in dams built on evaporites. The case of La Loteta Dam (NE Spain), a reservoir in a large karstic depression generated by interstratal salt dissolution. Engineering Geology, 185:139–154.

[6] Li J. Zhou W. (1999) Subsidence in karst mining areas in China and some methods of prevention. Engineering Geology, 52:45–50.

[7] Milanovic PT. (2000) Geological Engineering in Karst. Zebra, Belgrade. 347 pp.

[8] Song KI. Cho GC. Chang SB. (2012) Identification, remediation and analysis of karst sinkholes in the longest railroad tunnel in South Korea. Engineering Geology, 135–136:92–105.

[9] Waltham T. Bell F. Culshaw M. (2005) Sinkholes and Subsidence: Karst and Cavernous Rock in Engineering and Construction. Springer, Berlin. 382 pp.

[10] Youssef AM. Al-Harbi HM. Gutiérrez F. Zabramwi YA. Bulkhi AB. Zahrani SA. Bahamil AM. Zahrani AJ. Otaibi ZA. El-Haddad BA. (2016) Natural and human-induced sinkhole hazards in Saudi Arabia: distribution, investigation, causes and impacts. Hydrogeology Journal, 24(3):625–644. DOI: 10.1007/s10040-015-1336-0.

[11] Abdeltawab S. (2013) Karst limestone foundation geotechnical problems, detection and treatment: Case studies from Egypt and Saudi Arabia. International Journal of Scientific & Engineering Research, 4:376–387.

[12] Amin A. Bankher K. (1997a) Karst hazard assessment of Eastern Saudi Arabia. Natural Hazards, 15:21–30.

[13] Amin A. Bankher K. (1997b) Causes of land subsidence in the Kingdom of Saudi Arabia. Natural Hazards, 16:57–63.

[14] Edgell HS. (1990) Karst in northeastern Saudi Arabia. Journal of King Abdulaziz University, Earth Sciences, 3:81–94.

[15] Erol AO. (1989) Engineering geological considerations in a salt dome region surrounded by sabkha sediments, Saudi Arabia. Engineering Geology, 26:215–232.

[16] Kempe S. Dirks H. (2008) Layla Lakes, Saudi Arabia: The world-wide largest lacustrine gypsum tufas. Acta Carsologica, 37(1):1–14.

[17] Youssef AM. Al-Qaluobi H. Zabramawi YA. (2012a) Integration of remote sensing and electrical resistivity methods in sinkhole investigation in Saudi Arabia. Journal of Applied Geophysics, 87:28–39. Doi: 10.1016/j.jappgeo.2012.09.001.

[18] Youssef AM. Al-Qaluobi H. Zabramawi YA. (2012b) Sinkholes detection using electrical resistivity tomography (ERT) in Saudi Arabia. Journal of Geophysics and Engineering, 9:655–663. Doi: 10.1088/1742-2132/9/6/655.

[19] Pint JJ. (2003) The Desert Caves of Saudi Arabia: Stacey International, London, ISBN: 1 900988 48 8, 120 pp.

[20] Wittaker BN. Reddish DJ. (1989) Subsidence: Occurrence, Prediction and Control. Development in Geotechnical Engineering, 56, Elsevier, 528 pp.

[21] Davis SN. (1978) Origin of earth fissures. 74th Annual meeting, Tempe, Arizona, March 29–31, 1978. Geological Society of America, Abstracts, 10 (3), p 102.

[22] Helm DC. (1994) Hydraulic forces that play a role in generating fissures at depth. Bulletin Association Engineering Geologists, 31:293–304.

[23] Lofgren BE. (1978) Hydraulic stresses cause ground movement and fissures, Picacho, Arizona. Geological Society American Abstracts Programs, 10, p. 113.

[24] Roobol MJ. Shouman SA. Al-Solami AM. (1985) Earth tremors, ground fractures, and damage to buildings at Tabah (22/42C): Saudi Arabian Deputy Ministry for Mineral Resources, Technical Record DGMR-TR-05-4, 46 pp.

[25] Youssef AM. Maerz NH. (2013) Overview of some geological hazards in the Saudi Arabia. Environment and Earth Science, 70:3115–3130. Doi: 10.1007/s12665-013-2373-4.

[26] Holzer TL. (1984) Ground failure induced by ground-water withdrawal from unconsolidated sediment. In HolzerTL (Ed). Man- induced land subsidence, Rev. Eng. Geol., VI, the Geological Soc. of America, 67–105.

[27] Youssef AM. Sabtan AA. Maerz NH. Zabramawi YA. (2014) Earth Fissures in Wadi Najran, Kingdom of Saudi Arabia. Natural Hazards, 71:2013–2027. Doi: 10.1007/s11069-013-0991-5.

[28] Lund W. Lowe M. Bowman S. (2010) Land Subsidence and Earth Fissure Policy Recommendations, Cedar Valley, Iron County, Utah, Utah Geological Survey.

[29] Mousavi SM. Shamsai A. El Naggar MH. Khamehchian M. (2001) A GPS-based monitoring program of land subsidence due to groundwater withdrawal in Iran. Canadian Journal of Civil Engineering, 28:452–464.

[30] Sahu P. Sikdar PK. (2011) Threat of land subsidence in and around Kolkata City and East Kolkata Wetlands, West Bengal, India. Journal of Earth System Science, 120(3):435–446.

[31] Sun H. Grandstaff D. Shagam R. (1999) Land subsidence due to groundwater withdrawal: potential damage of subsidence and sea level rise in southern New Jersey, USA. Environmental Geology, 37:290–296.

[32] Wolkersdorfer Ch. Thiem G. (2006) Ground water withdrawal and land subsidence in Northeastern Saxony (Germaney), Mine Water and Environment Association.

[33] Xu YS. Shen SL. Cai ZY. Zhou GY. (2008) The state of land subsidence and prediction approaches due to groundwater withdrawal in China. Natural Hazards, 45(1):123–135.

[34] Zhang Y. Xue YQ. Wu JC. Shi XQ. Yu J. (2010) Excessive groundwater withdrawal and resultant land subsidence in the Su-Xi-Chang area, China. Environmental Earth Sciences, 61(6):1135–1143.

[35] Toufigh MM. Q`marsi K. (2002) Single well subsidence modeling based on finite element formulation, 2nd Canadian Specialty Conference on Computer Application in Geotechnique Proceeding, pp. 298–303.

[36] Amin AA.(1988) Potential Geologic Hazards at the Village of Tabah, Hail Region, MSc thesis, KAAU, Jeddah, 164 pp.

[37] Amin AA. Shehata WM.(1991) Subsidence due to groundwater withdrawal in a volcaniccrater, Presented in Hazards 91, Session 11, Perugia, Italy.

[38] Amin AA. Bankher K. (1995) Review on land subsidence in Saudi Arabia: in Barends, Brouwer and Schoder (Eds.), Land Subsidence, Balkema, Rotterdam, pp. 19–26.

[39] Bankher K. (1996) Engineering Geological Evaluation of Earth Fissures in Wadi Al-Yutamah South Al-Madinah Al-Munawwarah, M.Sc. Unpublished thesis, FES, Jeddah.

[40] Zabramawi YA. Youssef AM. Sabtan AA. Khiyami H. Bahamil AM. Al-Harbi HM. Hawsawi H. Zawahri M. Al-Wegdani K. (2009) Earth Fissures at Wadi Najran, Kingdom of Saudi Arabia: Saudi Geological Survey Technical Report SGS-TR-2009-12.

[41] Bahamil AM. Youssef AM. Khiyami H. Bulkhi AB. Otaibi ZA. Alshaikhi EA. (2016) Investigate and Evaluate of the Earth Fissures at Hail, Al Jouf, and Al Qasim regions (Al Qasim Region), Kingdom of Saudi Arabia: Saudi Geological Survey Technical Report SGS-TR-2016-4.

[42] Zabramawi YA. Youssef AM. Al-Harbi HM. Otaibi ZA. (2016) Investigate and evaluate of the Earth Fissures at Hail, Al Jouf, and Al Qasim regions (Hail Region), Kingdom of Saudi Arabia: Saudi Geological Survey Technical Report SGS-TR-2016-3.

[43] Zahrani SA. Youssef AM. Bulkhi AB. Al-Harbi HM. Otaibi ZA. Aamri M. (2016) Investigate and Evaluate of the Earth Fissures at Hail, Al Jouf, and Al Qasim regions (Al Jouf Region), Kingdom of Saudi Arabia: Saudi Geological Survey Technical Report SGS-TR-2016-2.

[44] Shehata W. Roobol JM. Stewart I. Khiyami H. Al Khammash A. Sayed S. Tarabolsi Y. Al Ahmadi K. (2007) Evaluation of a Subsidence Hazard in the Al Khafji area. Confidential report Saudi Geological Survey.

[45] Peters WD. Pint JJ. Kremla N. (1990) Karst landforms in the Kingdom of Saudi Arabia. The NSS Bulletin, 52:21–32.

[46] Vaslet D. Al-Muallem MS. Maddah SS. Brosse JM. Fourniguet J. Breton JP. Le Nindre YM. (1991) Geologic Map of the Ar Riyadh Quadrangle, Sheet 24 I, Scale 1:250,000, Kingdom of Saudi Arabia: Saudi Arabian Deputy Ministry for Mineral Resources Geoscience Map GM-121, 54 pp.

[47] Dini SM. Wallace CA. Halawani MA. Al-Sobhi SA. Kashghari WA. Al-Ghamdi AS. (2009) Geologic map of the as Sulayyil Quadrangle, Sheet 20H, Scale 1:250,000. Kingdom of Saudi Arabia. Saudi Geological Survey 64 pp.

[48] Memesh AM. Dini SM. Gutiérrez F. Wallace CA. (2008) Evidence of Large-Scale Subsidence Caused by Interstratal Karstification of Evaporites in the Interior Homocline of Central Saudi Arabia. European Geosciences Union General Assembly Geophysical Research Abstracts 10: A-02276.

[49] Dini SM. Al-Khattabi AF. Wallace CA. Banakhar AS. Al-Kaff MH. Al-Garni SM (2007) Geologic Map of Parts of the Judayyidet 'Ar'Ar Quadrangle, Sheet 31E, and Faydat al Adyan Quadrangle, Sheet 31F, Scale 1:250,000. Kingdom of Saudi Arabia. Saudi Geological Survey 45 pp.

[50] Al-Khattabi AF. Dini SM. Wallace CA. Banakhar AS. Al-Kaff MH. Al-Zahrani AM. (2010) Geologic Map of Parts of the 'Ar'Ar Quadrangle, Sheet 30E, Scale 1:250,000. Kingdom of Saudi Arabia. Saudi Geological Survey 50 pp.

[51] Gutiérrez F. Cooper AH. Johnson KS. (2008a) Identification, prediction and mitigation of sinkhole hazards in evaporite karst areas. Environmental Geology, 53:1007–1022.

[52] Gutiérrez F. Guerrero J. Lucha P. (2008b) A genetic classification of sinkholes illustrated from evaporite paleokarst exposures in Spain. Environmental Geology, 53:993–1006.

[53] Gutiérrez F. Parise M. De Waele J. Jourde H. (2014b) A review on natural and human-induced geohazards and impacts in karst. Earth Science Reviews, 138:61–88.

[54] Wallace CA. Dini SM. Al-Farasani AA. (2000b) Geological Map of the Wadi Sirhan Quadrangle, Sheet 30C, Kingdom of Saudi Arabia: Saudi Geological Survey Geoscience Map GM-127C, scale 1: 250,000.

[55] Banakhar AS. Al Zahrani AM. Al-Juaid AJ. Dini SM. Wallace CA. Al-Kaff MH. Al-Tassan AA. (2011) Geologic Map of the Markaz Adhfa Quadrangle, Sheet 29E, Scale 1:250,000. Kingdom of Saudi Arabia. Saudi Geological Survey, Geoscience Map GM-144C.

[56] Manivit J. Vaslet D. Berthiaux A. Le Strat P. Fourniguet J. (1986) Geological Map of the Buraydah Quadrangle, Sheet 26G. Kingdom of Saudi Arabia (with text): Saudi Arabia Deputy Ministry for Mineral Resources Geoscience Map GM-114C, scale 1: 250,000.

[57] Pint JJ. (2009) The Lava Caves of Khaybar, Saudi Arabia. Proceedings of the 15th International Congress of Speleology, Kerrville, Texas, 1873–1878.

[58] Powers RW. Ramirez LF. Redmond CD. Elberg JrFL. (1966) Geology of the Arabian Peninsula: Sedimentary Geology of Saudi Arabia. U.S. Geological Survey Professional Paper 560-D.

[59] Memesh AM. Dini SM. Al-Amoudi SA. Wallace CA. Sobhi SA. Al-Juaid AJ. (2010) Geologic Map of the Hawtat Bani Tamim Quadrangle, Sheet 21I, Sscale 1:250,000. Kingdom of Saudi Arabia. Saudi Geological Survey 64 pp.

[60] Zabramawi YA. Youssef AM. Qalioubi H. Memish A. (2011) Investigate and evaluate the karst problems along Al Atailiyah road, Kingdom of Saudi Arabia: Saudi Geological Survey, Technical Report SGS-TR-2011-1.

[61] Zabramawi YA. Youssef AM. Bahamil AM. Al Amri RM. Otaibi ZA. (2015) Investigate and Evaluate of the Karst Problems at Essawiyah Area. Kingdom of Saudi Arabia: Saudi Geological Survey, Technical Report SGS-TR-2015-11.

[62] Bulkhi AB. Youssef AM. Sabtan AA. Al-Harthi SG. Baamer MF. Al-Wegdani K. Al-Shereif A.(2009) Study and Surveying to Determine the Causes of Karstification in Al Aizizyah Area, Reyad, Kingdom of Saudi Arabia: Saudi Geological Survey Technical Report SGS-CR-2009-3.

[63] Carbonel D. Rodríguez-Tribaldos V. Gutiérrez F. Galve JP. Guerrero J. Zaroca M. Roqué C. Linares R. McCalpin JP. Acosta E. (2015) Investigating a damaging buried sinkhole cluster in an urban area integrating multiple techniques: geomorphological surveys, DInSAR, GPR, ERT, and trenching. Geomorphology, 229:3–16.

[64] Gutiérrez F. Galve JP. Lucha P. Castañeda C. Bonachea J. Guerrero J. (2011) Integrating geomorphological mapping, trenching, InSAR and GPR for the identification and characterization of sinkholes in the mantled evaporite karst of the Ebro Valley (NE Spain). Geomorphology, 134:144–156.

[65] McCalpin J. (2009) Paleoseismology, Second Edition. Academic Press. 629 pp. eBook ISBN: 9780080919980.

[66] Gutiérrez F. Galve JP. Lucha P. Bonachea J. Jordá L. Jordá R. (2009) Investigation of a large collapse sinkhole affecting a multi-storey building by means of geophysics and the trenching technique (Zaragoza city, NE Spain). Environmental Geology, 58:1107–1122.

[67] Gutiérrez F. Lucha P. Galve JP. (2010) Reconstructing the geochronological evolution of large landslides by means of the trenching technique in the Yesa Reservoir (Spanish Pyrenees). Geomorphology, 124:124–136.

[68] Carbonel D. Rosdríguez V. Gutiérrez F. McCalpin JP. Linares R. Roqué C. Zarroca M. Guerrero J. (2014) Sinkhole characterisation combining trenching, Ground Penetrating

Radar (GPR) and Electrical Resistivity Tomography (ERT). Earth Surface Processes and Landforms, 39:214–227.

[69] Griffiths DH. Barker RD. (1993) Two-dimensional resistivity imaging and modeling in areas of complex geology. Journal of Applied Geophysics, 29:211–226. Doi: 10.1016/0926-9851(93)90005-J.

[70] Gambetta M. Armadillo E. Carmisciano C. Stefanelli P. Cocchi L. Tontini FC. (2011) Determining geophysical properties of a near surface cave through integrated micro-gravity vertical gradient and electrical resistivity tomography measurements. Journal of Cave and Karst Studies, 73(1):11–15. Doi: 10.4311/jcks2009ex0091.

[71] Lambert DW. (1997) Dipole-Dipole D.C. Resistivity Surveying for Exploration of Karst Features. The Engineering Geology and Hydrogeology of Karst Terranes, Proceedings of the Sixth Multidisciplinary Conference on Sinkholes and the Engineering and Environmental Impacts of Karst, Springfield, Missouri, April 6–9. pp. 413–418.

[72] Roth MJS. Nyquist JE. (2003) Evaluation of Multi-Electrode Earth Resistivity Testing in Karst. Geotechnical Testing Journal, ASTM, 26:167–178.

[73] Van Schoor M. (2002) Detection of sinkholes using 2D electrical resistivity imaging. Journal of Applied Geophysics, 50:393–399.

[74] Zhou W. Beck BF. Adams AL. (2002) Effective electrode array in mapping karst hazards in electrical resistivity tomography. Environmental Geology, 42:922–928. Doi: 10.1007/s00254-002-0594-z.

[75] Zabramawi YA. Khiyami H. Youssef AM. Sabtan AA. Al-Harthi SG. Memash A. Al-Sharef A. Al-Ahmadi KZ. (2009) Karst Investigation and Analysis in An Nu'ayriyah Area, Kingdom of Saudi Arabia. Saudi Geological Survey Technical Report SGS-TR-2009-6.

Section 2

Indirect Manmade Geohazards through Climate Change

Case-Based Reasoning of Man-Made Geohazards Induced by Rainfall on Transportation Systems

Kuan-Tsung Chang, Edward Wang, Chih-Ping Kuo,
Min-Cheng Teng and James Jin-King Liu

Additional information is available at the end of the chapter

http://dx.doi.org/10.5772/64412

Abstract

Due to global warming and environmental change, disastrous natural events have increased in scale and impact, e.g., Typhoon Morakot, in 2009 and 2011 Tōhoku earthquake and resulting tsunami in Japan. Hazard management is becoming increasingly important, making it a necessity to manage risk and fully understand critical scenarios. For example, the National Infrastructure Protection Plan of the United States emphasizes on lessons learned from past disasters. In this chapter, several selected cases of accidents caused by man-made geohazards in Taiwan are studied.

The International Federation of Red Cross and Red Crescent Societies (IFRC) have identified technological or man-made hazards as events that are caused by humans and occur in or close to human settlements. These can include environmental degradation, pollution, and accidents (such as industrial and transport incidents). Accidents due to man-made hazards usually take place suddenly and give very limited time for response for rescue and recovery of function of facilities. Transportation facilities are a typical case. In this chapter, three hazard case studies are considered: a highway in southern Taiwan, a freeway in northern Taiwan, and an airport runway in the Taoyuan International Airport. The causes and the impacts of the incidents are described. These provide valuable lessons for managing this type of man-made hazard.

Keywords: infrastructure development, compound disasters, case-based reasoning, remote sensing, UAS

1. Introduction

1.1. Background

Climate change is a serious threat that could undo decades of infrastructure development in developing countries. While climate change is a global phenomenon, human activities are altering the local environment and will continue to do so. The Intergovernmental Panel on Climate Change (IPCC) established by the World Meteorological Organization (WMO) and United Nations Environment Programme (UNEP) has concluded that, over the past century, surface temperatures have increased, and associated impacts on physical and biological systems are increasingly being observed [1]. As climate change is altering rainfall patterns worldwide, scientists predict that wet areas will get wetter, dry areas will become drier, and storm tracks will move toward the poles [2]. Intensive rainfall has resulted in extreme flooding and landslides in many parts of the world. Floods in river basins have become the worst natural disaster, causing casualties, leaving people homeless, and disrupting transportation and economic activities. Floods have buried farmland and destroyed homes, factories, railroads, and bridges. Heavy rainfalls also triggered massive landslides.

The International Federation of Red Cross and Red Crescent Societies (IFRC) gave specific definitions on technological or man-made hazards in recent years. Events that are caused by humans and occur in or close to human settlements include environmental degradation, pollution, and accidents. Typical technological or man-made hazards are complex emergencies, conflicts, famine, displaced populations, industrial incidents, and transport accidents. They are all related to human habitat and modern civilization and can especially impact transport infrastructure such as highways and airports. Owners and operators of land transport systems exposed to rainfall-induced hazards are rarely aware of the risk-related concepts. This lack of knowledge affects the assessment of performance objectives and development of preventive measures for the sustainability of infrastructure systems with regard to flood events.

The majority of devastating disasters have occurred as a result of unusually heavy rains. Past events have highlighted the necessity to adjust the required design performance and specification level for new projects. However, these changes may not be cost-effective and require time to implement. Learning from past events can help facility owners and operators plan ahead regarding not only the exposure, but also the vulnerability and criticality of infrastructures. The design according to a specific return period, e.g., 100 years, may be appropriate for a new single infrastructure element. Assessing the impact of climate change on aging infrastructure can be difficult. Thus, the challenge engineers face today is not to control nature, but rather to adapt to it to lessen the adverse impacts of climate change. The wide range of lessons learned from past incidents can help establish a comprehensive approach addressing infrastructure security issues impacting the availability and quality of transport networks.

Engineers today apply Internet technology and remote sensing to provide unique solutions beyond what conventional methodology would normally provide. An unmanned aerial vehicle (UAV), for example, is an aircraft without a human pilot aboard used to perform

http://dx.doi.org/10.5772/64412

scientific observations and investigatory tasks. The UAV payload and flight stability has increased dramatically in recent years, utilizing spatial positioning components, such as Global Positioning System (GPS) and Inertial Measurement Unit (IMU), which are miniaturized to extend the flight time. UAVs also have the advantages of real-time wireless video transmission, low cost, flexibility, and low-level operations beneath clouds. These advantages can compensate for the shortcomings of conventional aerial or space remote sensing due to cloud cover, thus making UAVs an important aid for traditional aerial photogrammetry in obtaining spatial data. This technology enables engineers to learn and improve techniques from a new perspective.

1.2. Importance of this issue

A record high precipitation in southern Taiwan was set with 2,900 mm (114.17 inches or about 9.5 feet) of rain during Typhoon Morakot. On August 10, 2009, a single day record of 1,403 mm (55.23 inches) was set. According to Chris Burt's, "Extreme Weather" [3], the world record for 3-day rainfall is 127.56" on Reunion Island in 1952. Typhoon Morakot caused what officials claimed was the worst flooding in half a century. The number of known dead in Taiwan is 15, while 32 were severely injured. Those figures do not include landslide victims [4].

On April 25, 2010, a landslide occurred on a segment of the Formosa Freeway (Highway No. 3) near Xizhi. A large amount of dirt buried both directions of the freeway. Four cars were buried under the debris, killing four people. Bad hillside anchoring was blamed as a possible cause, as it had not been raining at the time of the collapse, and any earthquake had not been recorded [6].

In July 2014, underground gas lines exploded in the southern port city of Kaohsiung, killing 28. Heavy rainfall caused tremendous difficulties in the rescue efforts. On October 29, 2015, an EVA Airways Corp. aircraft sustained damage to its left horizontal stabilizer caused by the impact of a large piece of asphalt during takeoff on the southern runway at Taiwan's Taoyuan International Airport.

Although the facilities impacted in these cases are governed by different transportation authorities, scattered in various terrains with varying magnitude, the incidents share a common initiating event. This common factor is rainfall. Rainfall can be measured in the modern era with the global network of precipitation gauges. Surface coverage over oceans and remote areas is relatively sparse, but reducing reliance on interpolation, satellite clouds, and precipitation data have been available since the 1970s. Modern engineers may also take advantage of satellite imaging to develop solutions to mitigate problems.

The developing trends of disaster mitigation and management have focused on risk management through analyzing hypothetical scenarios. The National Infrastructure Protection Plan of the United States emphasizes lessons learned from past disasters. It is believed that the governing authorities and decision makers in disaster management organizations may reduce the impact of hazards and the vulnerability of society by utilizing past experience. Executive actions and preventative measures developed considering history may help reduce the potential loss of life and property.

1.3. Purpose of this chapter

The purpose of this chapter is to report three unexpected incidents in detail. All these cases impacted transportation infrastructure. Areas examined and addressed include the cause of the disaster; what the engineering solution brought to the problem; and how engineers in Taiwan can prevent a similar tragedy from happening again.

In addition, the utilization of data acquisition from high-resolution photo images obtained through the use of UAVs after the incident will be considered. This enables engineers to assess the current site conditions, perform safety evaluations, and plan mitigation procedures.

2. Case studies

The cases discussed are as follows:

- Collapse of Chung Lin Road in Kaohsiung's Siaogang District - subsidence cracked Chung Lin Road for the second time within a year in the aftermath of Typhoon Dujuan in September 2015.
- Landslide on Freeway No. 3 - the slope failed at the 3.1-km mark from the northern end, just north of the Chitu toll station, and collapsed an overpass on April 25, 2010.
- Taoyuan International Airport pavement cracks - Taiwan's largest airport opened on February 26, 1979. Rain seepage into the pavement caused excessive hydrostatic pressure during plane take-off.

2.1. Collapse of Chung Lin Road in Kaohsiung's Siaogang District

The collapse at Chung Lin Road occurred suddenly around 2:00 a.m. on 18th September 2015, in an area where shield tunnelling was being conducted. Due to the collapse, drinking water pipes were destroyed, and electricity was interrupted due to the tilting of power poles. In addition to the inconvenience for residents, nearby petrochemical pipelines were also endangered. This could have jeopardized the supply chains of petrochemical feedstock.

2.1.1. The location of the event

Figure 1 shows the location of the collapse of Chung Lin Road, Kaohsiung City. The collapse measured 40 m by 10 m. The site is close to the Dalin Refining plant of the Refining Business Division's Operators of The CPC Corporation, Taiwan. The plant includes a number of vulnerable facilities. The distance of the collapse to the nearby oil tanks is only a few 100 m. Fortunately, CPC executed an emergency stop of the transportation of the fluids through the pipelines, and thus no leakage or explosion was triggered.

http://dx.doi.org/10.5772/64412

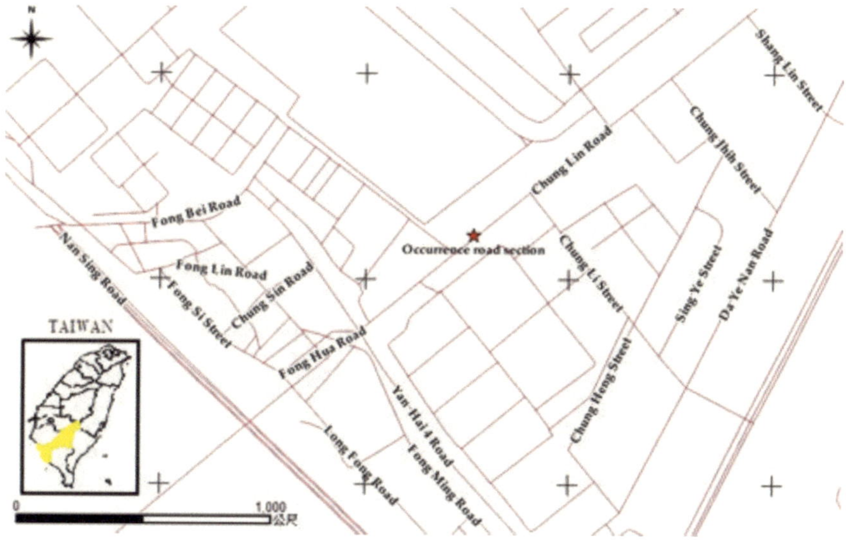

Figure 1. Location of the event.

2.1.2. *General description of the event*

2.1.2.1. *Introduction*

Between 2 a.m. and 6 a.m. on September 18, 2015, an accident occurred in the underground cable shield tunnelling engineering of Taiwan Power Company. It was located underneath the road from Chung Lin Road intersection to the direction of Talinpu in Hsiaokang District, Kaohsiung City and caused the road to collapse. In the early stage of the accident, the road collapse extended to about 40m long and about 10m wide (see **Figure 2**). Power poles tilted and caused an electrical outage. Due to safety considerations, the CPC Corporation suspended the use of 11 hazardous substance pipelines for hydrogen gas, natural gas, pure benzene, toluene, xylene, fuel oil, crude oil, refined oil product, etc. This closure forced the medium-carbon light-oil processing factory to shut down, which led to implementing production and marketing emergency-response measures to avoid affecting Taiwan's oil supply.

Preliminary estimates show that the direct property loss, resulting from this collapse accident, is about 500 million Yuan. This includes the plant facility damage of about 280 million Yuan to CPC Corporation and China Steel Corporation and 220 million Yuan of losses to other affected business and agencies (Kaohsiung Linhai Industrial Park, Sewage Treatment Plant of Linhai Industrial Park, Taiwan Water Corporation, Chunghwa Telecom Corporation, Water Resources Bureau, Kaohsiung City Government, etc.). It is not possible to accurately estimate the indirect losses such as production suspension at CPC Corporation (where the loss in daily capacity was close to 40 million Yuan) and the disruption of raw materials to downstream facilities.

Figure 2. Destruction of roadway.

2.1.2.2. Preliminary analysis of causes

The accident occurred during the underground cable tunneling project of Taiwan Power Company. Due to the failure of the shield tunneling engineering machinery occurring 30 m underground, groundwater flooded into the excavation and caused the road pavement to collapse. In addition to the underground drinking water pipe failures, trees, and utility poles around the road tilted as well. Although the accident occurred at about 2 a.m. on the 18th of September, the construction personnel did not formally close the road and notify the impacted pipeline and business units until past 6 a.m. In addition, since the underground pipelines were already damaged, secondary damage occurred during the emergency repair, including oil and water leakage with the resulting environmental pollution. The collapse area continued to expand after the initial accident.

2.1.3. Social impacts

This incident highlights the importance of underground pipeline safety management. The pipeline was buried before the urban area developed. The side-by-side development of the metropolitan and industrial areas was inevitable due to the limitations of land use. Therefore, the risks to public safety from accidents such as pipeline leakage, explosion from the hazardous substance, and public pipelines buried underneath roads are increased, once these facilities are damaged by the excavation accident.

It is fortunate that this incident occurred early in the morning during the off-peak traffic period. Although there were no casualties, it caused the road collapse and closure, direct losses to the state business, impacts onto the petrochemical industry operation, and subsequent inconvenience to peoples' lives. The impacts caused by this incident are as follows.

2.1.3.1. *Interruption of road passage*

In the accident area, the large-scale collapse destroyed Chung Lin Road, causing its closure, creating inconvenience for local motorists. Fortunately, an emergency road, allowing temporary motor-vehicle access, was completed about 2 months (November 15, 2015) after the accident.

2.1.3.2. *Interruption of lifeline infrastructure*

Due to the severity of the collapse, the infrastructure components around the road were all seriously damaged. Utility infrastructures, like power-supply equipment, and communication and public pipelines are also damaged, disrupting the water, power, and gas supplies, interrupting telecommunications and severely affecting the residents' normal life.

2.1.3.3. *Impact of industry*

Because the CPC Corporation closed the underground petrochemical pipelines underneath the accident area, the affected petrochemical industry facilities were forced to face the crisis of a production shutdown. Petrochemical raw materials would have to be transported with tank vehicles or alternative pipelines, thus increasing the risk of hazardous-substance transportation incidents on the road.

2.1.4. *Lessons learned from the event*

The aforementioned accident can provide the following lessons.

1. Identifying the risks of the construction method selected beforehand would have identified potential accident consequences.

2. Immediate communication with local authorities and industries could have reduced the danger to the public and oil-related pollution.

3. Development of a coordinated contingency plan, in advance, with emergency responders could have reduced the impact and loss caused by the disaster.

2.2. Landslide on Freeway No. 3

Section-3.1K landslide accident of Freeway No. 3 occurred at Mt. Shihgongge at Location 3.25K at 14:29 p.m. on April 25, 2010. According to statistics, this is the most serious landslide accident in the past 36 years on Taiwan's national freeway. The worst landslide on the system occurred near Badu Interchange on the Sun Yat-sen Freeway on September 28, 1974, which caused 36 deaths [5–8].

2.2.1. The location of the event

The landslide accident occurred in Madong Village, Qidu District, Keelung City. During the slide, the Dabu Bridge crossing Freeway No. 3 also collapsed onto the freeway mainline, thus blocking traffic (**Figure 3**).

Figure 3. Photograph of the landslide at Freeway No.3 (Source: http://www.nasc.gov.tw/files/duty_news_files/_4251883.JPG).

The accident took place at a dip-slope, with the site elevation varying between 122 m to 182 m above sea level. The dip angle of the strata at this location ranges from 12 to 15°. The geological formations consisted of Daliao Shale and sandstone as well as shale alternations. The lithology of the sliding surface is on a shale formation.

2.2.2. General description of the event

2.2.2.1. Introduction

In this large-scale landslide accident, four deaths occurred. The day the event occurred was sunny and free of typhoon, rainstorm, earthquake, and other factors that typically cause landslides. According to subsequent analysis, the possible cause was that, due to side-slope excavation of freeway engineering, the free end of the dip slope toe was exposed (or daylighted). The rock material is block sandstone with shale inter-bedding. A weak surface is likely to be caused between beds. Underground water infiltrates in the sandstone (which has high permeability), resulting in hydrolyzing and weathering on the top of the thick shale, and finally causing sliding failure of weak surface. According to the comparison of topographic data before the landslide disaster (obtained through Airborne LiDAR in April 2006) and the measured data after the landside disaster, the collapsed earth and stone volume is 165,000 m^3, the collapse area is about 1.14 hectares, the influenced scope is 2.4 hectares, and the collapse depth is 15 to 20 m.

2.2.2.2. Preliminary analysis of causes

The site is a mostly oblique ridge with a steep slope in the northwest and gentle slope in the southeast in this region, and it is dip slope in terrain, that is, the stratum tendency and slope direction are close, and in the southeast direction. There are many similar dip slopes along the alignment of the freeway in northern Taiwan. According to the field investigation results, the landslide causes are preliminarily summarized as follows.

1. The disaster was caused by dip-slope sliding.
2. The dip slope at the location is formed through mutual stacking of sandstone and shale, and a slide is likely to occur at the interface of rock layers.
3. The sandstone in the upper layer is seriously weathered, rich in vertical joints, and permeable. The shale in the lower layer has poor water permeability; therefore, a higher water content and water pressure was caused in the interface between the two rock layers, reducing the friction between the rock layers and resulting in a slide.
4. The sliding rock is huge, and it is likely to exceed the original rock anchor capacity, causing the breakage of the sliding rock's connection to the base material.

2.2.3. Social impacts

2.2.3.1. Interruption of road passage

The landslide covered six lanes with earth and stone. Three cars were buried, causing four deaths. The Dabu Bridge broke into two sections and fell onto the freeway mainline. Therefore, National Freeway Bureau and National Highway Police Bureau emergently blocked the section from the site north to Sijhih System Interchange of Freeway No.3, and from the site south to Keejin Interchange, leaving traffic flow completely disconnected.

2.2.3.2. Interruption of lifelines

The traffic shutdown lasted for more than 70 days until the accident section was reopened (June 19, 2010). Commuters in Keelung and other regions had to select alternative routes due to the closure of the collapsed section, adding 1.5 hours for the journey from Taipei to Keelung. Some traffic detoured to the Sun Yat-sen Freeway or provincial highways, causing traffic jams in many locations. People's normal lives were seriously impacted by the closure.

2.2.3.3. Impacts on industry

The north-south two-way section from Keelung to Taipei of the Sun Yat-sen Freeway suffered traffic jams in rush hours. This added significant transportation time and cost to the import and export traffic from Keelung port.

2.2.4. Lessons learned from the event

After the landslide accident, the Department of Transportation organized an academic unit to carry out "investigation work of 3.1K landslide accident of Freeway No.3" to identify the accident causes and study subsequent emergency treatment strategies. The major findings of the investigation are as follows [6–8].

There is a difference in the included angle between the original design section used in the side-slope stability analysis and the actual sliding section. According to the side-slope stability analysis code at the time, the safety factor upon earthquake should be bigger than or equal to 1.1. The calculated safety factor of the original design section is in compliance with the requirements given in the code. However, the calculated safety factor of the actual sliding section is only 0.96 and is not in compliance with the requirements given in the code. Moreover, the side-slope stability safety factor of the actual sliding section under earthquake state is not in compliance with the requirements given in the code. These discrepancies between the original design and actual field-condition calculation results reflect the critical importance of field-surveying accuracy.

Ground anchors were used to restrain the section from sliding, but the anchor forces may have impacted the rock stress distribution and caused tension in the upper part of the slope. This is an area requiring further research and possible adjustment of the safety factors in the existing code.

Of the 572 ground anchors completed in the beginning of 1998, only 58 remained on the side slope after the landslide, with the damage ratio reaching as high as 90%. This reflects the importance of ground anchor and tie-back tensile force in side-slope maintenance.

In earlier inspections, the ground anchor-related investigation was not conducted. In addition, the regular inspection reports were not filled out in detail. Therefore, it is not known whether the pre-stress of the ground anchors gradually relaxed or the tendons of ground anchors were rusted in the period from the time the ground anchors were completed (at the beginning of 1998) to the time landslide occurred. This indicates the importance of regular comprehensive inspection.

The governing authority ignored the collapse risk as well as neglecting the importance of the existing monitoring equipment, so that the use of the landslide detector was stopped without prudent consideration of the potential consequences, indicating the importance of automatic, continuous monitoring.

In terms of treatment and maintenance of soil and water conservation, the slope stability and water-drainage systems should be "considered as a whole." In future, therefore, for the indivisible places of upper and lower side slopes of highways, the highway administration authority will be specified to be responsible for maintenance of the highway-facility safety. The highway authority will appoint personnel to enter into the public and private land to inspect and test conditions; and the land owner, user, or manager shall not evade, hinder, or refuse this inspection [8].

In future, further research has to be undertaken on whether tension cracks occur in the upper part of the dip slope outside the highway right-of-way impacting the ground anchors, similar to those on the collapsed side slope in this case. This could cause surface water leaking to rock stratum, thus weakening the strength between rock strata, resulting in similar effects as the collapse at this site [8].

2.3. Taoyuan International Airport Pavement Debris

On Thursday, October 29, 2015, an aircraft owned by Taiwan's second largest carrier, EVA Airways Corporation, sustained damage to its left horizontal stabilizer during take-off on the southern runway at the Taoyuan International Airport. This was caused by the impact of a large piece of asphalt from the runway with the stabilizer.

Asphalt blowup in an airport runway is very dangerous for aircraft operations and requires immediate attention. Airport pavement distresses begin with minor moisture leakage, which, when deemed detrimental to aircraft operations, can lead to runway closure. The impact of the runway closure not only delivered a negative message to other countries but also caused a huge loss to the tourism industry.

2.3.1. The location of the event

The Taoyuan International Airport serves the capital city of Taipei, Taiwan, and the northern parts of the island. Located about 40 km west of Taipei, the facility is Taiwan's largest airport with regular international flights. It is by far the busiest international air entry point in Taiwan and the main international hub for China Airlines and EVA Air. The airport opened for commercial operations in 1979 and is an important regional trans-shipment center, passenger hub, and gateway for destinations in Asia. The airport was formerly known as Chiang Kai-shek International Airport (CKS International Airport), in remembrance of the former president, until the name was changed on 6 September 2006. The number of aircrafts landing and departing from the airport has grown from 150,000 to 200,000 per year. The Taoyuan International Airport handled a total of 35,804,465 passengers and 2,088,726,700 kg of freight in 2014. It is the 11th busiest airport worldwide in terms of international passenger numbers and fifth busiest in terms of international freight traffic.

2.3.2. General description of the damage

The damage incurred to the EVA Air jet was alleged to cost EVA Airways Corporation close to $10 million Yuan (US$ 300,000). The runway maintenance problem at the airport has not only caused flight delays, but the hazard potential has also threatened aviation safety. The original airfield pavements were designed and constructed to provide an adequate support for the various loads imposed by both aircraft and environmental (climate) conditions such as temperature or moisture variations. The runways were also constructed with joints to allow expansion and contraction from temperature changes. The initial investigation of the incident attributes it to hydraulic fluid leaking from aircraft causing erosion of the pavement. A further analysis suggested that the design of the runways was flawed, if that was the main cause of the damage. However, the Civil Aeronautics Administration and airlines denied that theory of erosion being caused by hydraulic fluids, saying that the detection of such a hydraulic-fluid leak would result in an aircraft being grounded for immediate repairs.

2.3.3. Social impacts

For airport pavement, distress can be caused by the disintegration of pavement due to axial compression forces generated by slab expansion due to pavement temperature and moisture changes. Blowup usually occurs at transverse joints or cracks in hot weather if they are not wide enough for pavement expansion. If pressure from pavement expansion cannot be relieved, it results in a localized upward movement of slab edges or shattering in the vicinity of the joint [9]. As reported in the Central Region Airport Certification Bulletin, an airport runway pavement blowup case occurred at the Ankeny Regional Airport in Iowa, United States, in the summer of 2011. A similar case occurred in Nepal where a number of international flights in the Tribhuvan International Airport in Kathmandu Nepal were delayed, diverted, and cancelled due to airport flexible-pavement distress during August 2013. Pavement-related failure incidence around the globe shows that pavement deterioration, pavement debris, and mud blowup as major airport safety concerns are common phenomena in airports.

Pavement distress and the resulting Foreign Object Debris (FOD) (a term referring to a foreign substance or debris that can cause aircraft damage) on airport runways are dangerous for aircraft operations. If FOD occurs on airport runways without prompt discovery and removal, an aircraft can be damaged during takeoff or landing. A serious accident can result, and consequently the passengers aboard the aircraft may be injured or may lose their lives. The runway needs to be closed for repairing distress and removal of FOD. A closed runway causes economic losses resulting from flight delays, cancellations, etc. To avoid the recurrence of similar incidents, the Taoyuan International Airport Corporation (TIAC) has changed the pavement at both ends of the runway from asphalt to cement, adding that the firm plans to procure an FOD detection system—the same as the one used at the Hong Kong International Airport—to monitor runway conditions at Taoyuan Airport.

2.3.4. Lessons learned from the event

Pavement health monitoring employing advanced technology to allow the assessment of the structural reliability and detection of structural changes is an effective solution to prevent aircraft accident and damage caused by poor pavement performance and FOD. However, the full solution of the airfield pavement debris problem should start by analyzing its fundamental causes. An investigation of the compacted backfill under the airport runway slab using ground-penetrating radar (GPR) showed substantial voids. In addition, after a continuous rainfall, a great deal of moisture seepage occurred through the construction joints. Under a cyclic surface loading (resulting from airplane takeoff and landing), the hydrostatic pressure generated by the noncompressible fluid washed away the foundation substrate (from small to mid-size soil particles) leaving more spaces for additional water. The negative cycle continues until—when there are not enough voids to accommodate or relieve the pressure—the slab cracks occur.

To resolve the particular issue, Taiwan engineers have gone through a series of laboratory tests to imitate the phenomena in a controlled environment. A pilot program implementing Hydrostatic Pressure Relief Technology has been proposed and is under construction. The idea came from the pressure-relief mechanism used under railways in Europe where a permeable geosynthetic layer is installed underneath the slab. The Hydrostatic Pressure Relief Technology has also been adopted successfully by developers and building contractors in more than 60 cases where buildings are subjected to underground water buoyancy. In addition, through a continuous remote monitoring and review of settlement, seepage, and pore-water pressure, the proposed system is expected to provide a rapid, simple, and cost-effective solution to the airfield pavement-deterioration problem.

3. Topographic change analysis of man-made geohazards with unmanned aerial vehicles

An unmanned aerial vehicle (UAV) (also known as a drone) is an aircraft without a human pilot aboard used to perform scientific observations and investigation tasks. Since UAV payload and flight stability have recently increased dramatically, spatial positioning components, such as GPS and IMU, have been miniaturized to extend the flight time. UAVs also have the advantages of real-time wireless video transmission, low cost, flexibility, and low-level operations under the cloud base [10]. These properties can compensate for conventional aerial or space remote sensing subject to the shortcomings of cloud cover; hence, it becomes one of the important aids for traditional aerial photogrammetry to obtain spatial data. Many kinds of sensors, e.g., hyper-spectrometer, thermal imager, or LiDAR sensors, can be mounted on a UAV, although the most commonly used sensor is the consumer-grade digital camera [11].

The UAV has been widely used in the world, including maritime search and rescue, forest conservation, soil and water conservation, natural disaster investigation, and so on [12–19].

The authors used an eBee UAV for collecting high-resolution geospatial data for golf-course maintenance [20]. The study showed that the UAV was efficient and of low cost for producing orthophoto of 3-cm resolution and 3-D terrain point clouds for a golf course. The benefits derived from geoinformatics in the maintenance and management of a golf course had also been demonstrated. Since there was no spatial data as a map database for the golf course, the results indicated that the very high-resolution 3-D spatial information can provide an important database to assist the facility staff in performing improvement planning and design, utilizing spatial analysis tools. Moreover, golf players can access the spatial information services using spatial query functions to learn where various features and amenities are located. This could provide course managers an excellent marketing tool for providing golf players with better value-added services.

Another UAV study was conducted for a case of drainage planning. For mapping the efficiency and higher visual communication for drainage planning, a combination of aerial survey with unmanned aerial vehicles and a supplementary field survey was conducted to produce very high-resolution orthogonal-rectified images and 3-D terrain data in the sludge sedimentation tank area of the Shihmen Reservoir, northern Taiwan. The result of accuracy evaluation with 25 check points shows that the mean errors in X, Y, and Z are 7.57 cm, 8.36 cm, and 23.4 cm, respectively. According to the topographic mapping standard in the scale of 1/1000, the mean error is acceptable for aerial mapping [21]. The abovementioned cases showed that UAVs can be effectively applied to collect information for the assessment of structure deformations and damage when man-made geohazards take place. The results can be used for strengthening the structures or developing other measures for mitigating hazards.

3.1. Procedures of data acquisition with a UAV

First, a flight plan can be made based on the requirements of the emergency mission and the sensor system used, including items such as the specifications of the camera used, overlap ratio of photos, and the condition of the disturbed ground surface. UAVs come in many shapes and sizes. The characteristics or specifications of the UAV ultimately lead to the operator's decision as to which platform will best fit the survey application. These key attributes and acting on them will ensure that the mapping mission is a success. Two main types of UAV are available that are suitable for surveying work. The first type is a fixed-wing model. In general, the stability of a fixed-wing UAV is good for precision mapping for topography, but it does have certain restrictions in taking-off field and operation conditions. The second type of UAV is a rotary-blade, or propeller-based model. Unlike the fixed-wing models, these mini-copters are able to fly in every direction, horizontally and vertically, as well as hover in a fixed position. This makes them the perfect instrument for detailed inspection work or surveying hard-to-reach areas such as pipelines, bridges, power lines, and rail tracks. A fixed-wing UAV can be suitable for a wide area survey, whereas a rotary UAV can be better adopted for complicated terrain and restricted open space [22].

Secondly, control points and check points should be properly selected to cover the survey areas. The quantity of the required points depends on the quality of the final outputs required. If they are digital terrain models and orthophotos of high quality, this will require high quality of the control points. Otherwise, a few points (such as 3 points) can be enough for a reconnaissance. Subsequently, the geodetic coordinates of the control points can be measured by Real-Time Kinematic (RTK) satellite navigation or measured with a Control Entity Database using control entities of aerial images. In addition, weather conditions are critical for image quality due to the illumination and vibration due to wind speed. Before an aerial sortie, all functions of the platform and sensor should be carefully checked.

After imaging, all captured images are imported into a postprocessing software such as Pix4D, AgisoftPhotoScan, or Postflight Terra 3D software to perform photogrammetric processing of digital images and to generate 3-D spatial data. For example, it is possible to apply Agisoft-PhotoScan for a full automation of the postprocessing, including image registration for orientations and digital elevation modelling, with either traditional aerial photographs or digital images [23]. An SfM (Structure from Motion) algorithm is adopted in the software to accommodate the situation when there are no control points and/or inner parameters of the camera. SfM will retrieve the spatial coordinates of objects in the stereo-pairs by reconstructing the location and orientations of all the images captured in the air. The procedures of SfM approach include:

1. Reconstruction of the locations and orientations of each stereo-pair;

2. Construction of the trajectories of the camera; and

3. Construction of the 3-D landscape.

Generally, the payload capability of UAVs is very limited. Consumer digital cameras are used to replace the conventional metric camera. Thus, it is a prerequisite to calibrate the digital camera to obtain its parameters, including the inner orientation parameters and distortions. These parameters are entered into the postprocessing software. Feature points and conjugate points can be retrieved in the software with a function such as AlignPhotos. Subsequently, photo centers and orientations of each exposure can be derived. The second stage of the procedure is to use the function of Build Sense Cloud to generate point clouds on the basis of a bundle adjustment and ray-tracing approach. This is optimized with filtering out outliers. The third stage of the procedure utilizes the function of Build Mesh to construct a TIN (Triangular Irregular Network) for the point clouds generated in the previous stage. The digital surface models (DSMs) can be generated. The fourth stage of the procedure is applying the function of Build Texture to construct the texture of the 3D models by draping corresponding textures expressed on aerial photographs. It is noteworthy that this involves the reconstruction of the camera parameters of location and orientations. These parameters are better entered as extra parameters for compensating the automatic reconstruction of the adjustment of aerial triangulations and inner orientations [24]. Accurate stereo-models can thus be established with the exterior orientation of each camera exposure stations. Digital Terrain Models (DTMs) and ortho-photographs can be automatically generated subsequently. Finally, animated videos of the geohazards, the 3D models, and other useful thematic maps such as slope gradient, slope aspect, and contour lines can be derived as well for other geospatial analyses such as earth-volume analysis with Difference of DTMs (DoD) and for assisting in damage assessment and planning mitigation measures.

3.2. Suao Landslide Case at the Coastal Su-Hua Section of Highway Route 9

Due to the combined effects of Northeast seasonal winds and Typhoon Nalgae, heavy rainfall induced in the I-Lan area resulted in serious landslides. The Suao Landslide at the coastal Su-Hua Section of Highway Route 9 was located at the 11.8K-km mark. At this site, the entrainment effect of debris flow and toe erosion on the downslope induced a regressive sliding failure at the adjacent road causing the downslope collapse and roadside barrier failure. The rockfall barriers at the upper stream of the nearby Daken bridge were destroyed as well. The incident took place at 05:00 on 9th October 2011. The heavy rainfall lasted 12 h, with an average rainfall at the nearby rain station of 30.5 mm/h with a total rainfall of 418.9 mm. The affected landslide area along the road was 150 m^2. **Figures 4** and **5** show the photographs captured by the UAV camera. **Figure 4** shows the orthorectified image of the landslide. **Figure 5** shows some selected views of the landslide.

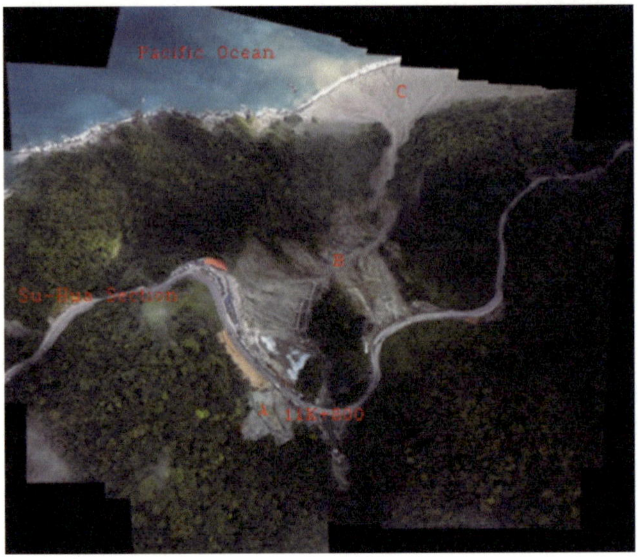

Figure 4. The ortho-rectified UAV image of the Suao Landslides (Source: GangYu Corporation).

Figure 5. Some selected views of UAV images of the Suao Landslides (Source: GangYu Corporation).

3.3. A Dip-slope Landslide on Freeway No. 3

A dip-slope landslide on Freeway No. 3, close to Chidu District of Keelung City, northern Taiwan took place on 25th April 2010. The landslide has an area of 11,475 m^2. UAV was applied as an experiment of emergency response to acquire aerial images of the affected area, to produce stereo-models, and to evaluate the damage for the analyses and interpretation of the disaster [25].

The procedures of UAV operation were as follows. Image acquisition was conducted on 26th April 2010, the day after the event. There was a clear sky with light haze. Flight height was 500 m above ground in average. Flight speed was 50 km/h. Ground resolution of the images was about 17 cm. The forward overlap between images is more than 80%. **Figure 6** is a UAV side view of the landslide captured by a rotary-wing UAV owned by the Flying Tiger Helicopter and Skyline Dynamics Co., Ltd. **Figure 7** is a historical photograph of the site taken by the Agricultural and Forestry Aerial Survey Institute (AFASI).

Figure 6. UAV side-view of the landslide taken by a rotary wing UAV owned by Flying Tiger Helicopter and Skyline Dynamics Co., Ltd.

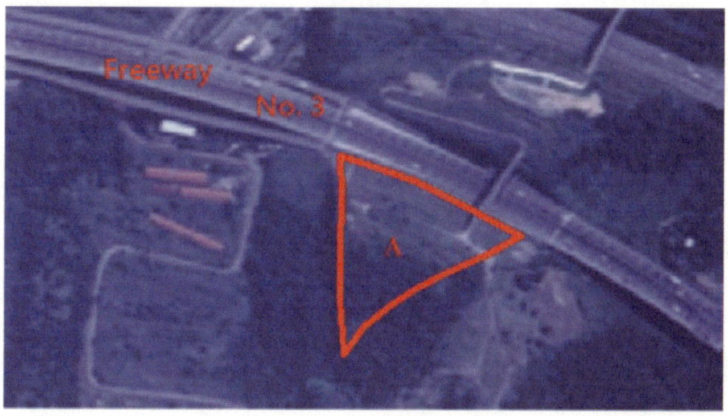

Figure 7. Historical photograph of the site taken by the Agricultural and Forestry Aerial Survey Institute (AFASI).

To facilitate the production of stereo-pairs, the stereo-models developed for producing official electrical maps (a scale of 1 to 2500) in 2009 were used for measuring control points of ground features such as building corners and road signs. In total, one vertical control point and three full control points were captured along with three check points for subsequent accuracy assessment. In a later stage, image automatic matching and aerial triangulation were conducted with Erdas LPS and 25 tie points. **Figure 8** shows the locations of all the control points, check points, and tie points.

After the establishment of stereo-pairs, digital elevation models and orthophotomosaics were created. **Figure 9** shows the side boundary of the landslide and the limit of the freeway where the landslide overtopped it. The landslide area is estimated as 11,475 m^2 and the volume 137,571 m^3, with an estimated material weight of 300,000 tons. These estimates were useful for managing the trucks used for debris disposal.

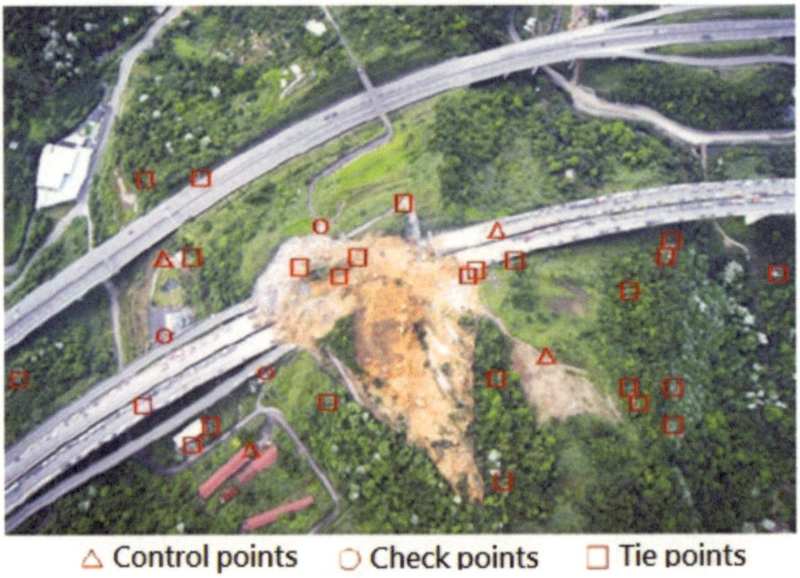

Figure 8. Locations of all control points, check points, and tie points [25].

Figure 9. Side boundaries of the landslide and the freeway limits [23].

Additional benefits that the decision makers can gain from the results of UAV operations include 3-D schematic models of the landslide (**Figure 10** and **Figure 11**), animated videos, topographic analysis of the landslide, and a comparison of the cross sections prior to and following the event (**Figure 12** and **Figure 13**). To facilitate the process of emergency response, a standard procedure, including UAV operations, can be beneficial and worthy of establishment.

Figure 10. 3-D Models prior to the event [23].

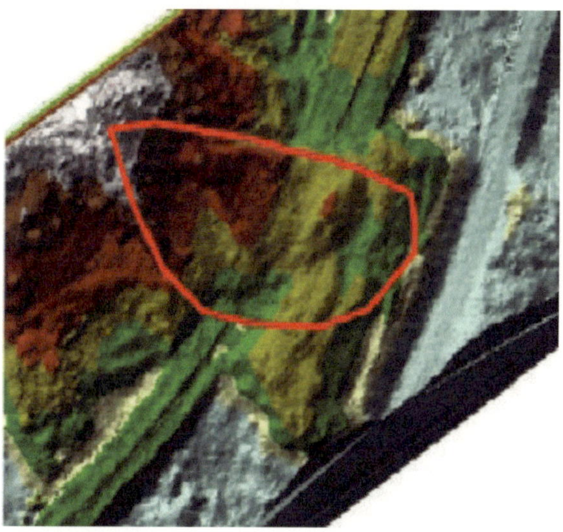

Figure 11. 3-D models following the event [23].

Figure 12. Location of the cross section.

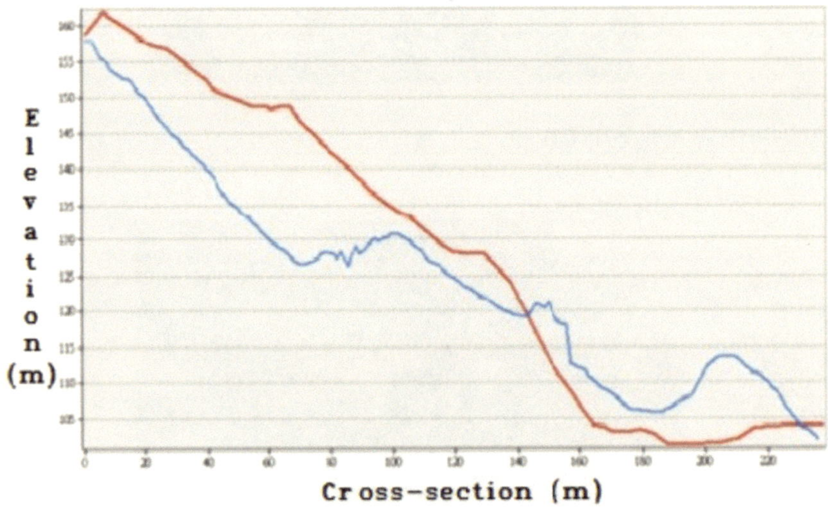

Figure 13. Cross sections for both prior to and following the event.

4. Conclusions and suggestions

The results of the man-made geohazards in the three case studies considered can be summarized as follows:

1. To minimize the impacts and losses due to disasters, information sharing for emergency responses is critical. In addition, preparedness and monitoring of operations are required before any emergency cases occur.

2. The case of Freeway No. 3 showed that the slope stability did not meet the code requirements for earthquake resistance. This was due to the uncertainty of the in situ measurements. Therefore, the accuracy standards, the design parameters for ground anchors, the procedures of regular inspection, and automated sensing for certain vulnerable zones are all important to prevent similar future events.

3. The Hydrostatic Pressure Relief Technology has been adopted successfully by developers and building contractors in more than 60 cases where buildings are subjected to underground water buoyancy. In addition, through a continuous remote monitoring and review of settlement, seepage, and pore-water pressure, the proposed system is expected to provide a rapid, simple, and cost-effective solution to airfield pavement-deterioration problems.

4. UAVs equipped with suitable sensors will become a critical means for assuring the process of design, construction, operation, and emergency response for incidents caused by man-made geohazards. UAVs can be used for quick generation of the DTM, orthophotos, 3-D models, animated birds' eye view, and change detection for assessing hazards. The procedures of UAV applications and productions of 3-D information are presented in this chapter.

Acknowledgements

The authors would like to thank GangYu Corporation, Flying Tiger Helicopter, and Skyline Dynamics Co., Ltd. for the use of their UAV images.

Author details

Kuan-Tsung Chang[1*], Edward Wang[1], Chih-Ping Kuo[1], Min-Cheng Teng[2] and James Jin-King Liu[3]

*Address all correspondence to: ktchang1216@gmail.com

1 Department of Civil Engineering and Environmental Informatics, Minghsin University of Science and Technology, Hsinfong, Hsinchu, Taiwan

2 National Science and Technology Center for Disaster Reduction, New Taipei City, Taiwan

3 LiDAR Technology Co., Ltd., Zhubei, Hsinchu, Taiwan

References

[1] IPCC. Poverty and climate change – reducing the vulnerability of the poor through adaptation. In: Eigth Conference of Parties to the United Nations Framework Convention on Climate Change; 2002; New Delhi: OECD; 2002. Available from: www.oecd.org/env/cc/2502872.pdf [Accessed: Jan. 2016].

[2] Kate M., Céline B. Identifying external influences on global precipitation. Journal Proceedings of the National Academy of Science. 2013; 110(48):19301–19306. DOI: 10.1073/pnas.1314382110.

[3] Environmental Protection Administration. Extreme events and disasters are the biggest threats in Taiwan [Internet]. 2009 [Updated: 2009]. Available from: http://unfccc.epa.gov.tw/UNFCCC/ [Accessed: Jan. 2016].

[4] Zhao Y. Key technologies in low-cost integrated vehicle navigation systems [dissertation]. Stochholm, Sweden: Royal Institute of Technology (KTH); 2013. Available from: https://www.diva-portal.org/smash/get/diva2:656086/FULLTEXT01.pdf.

[5] NCDR. Preliminary report on the survey and analysis of road-side slope failure at the 3.1-kilometer mark of Freeway No. 3 [Internet]. 2010 (in Chinese).

[6] Taiwan Geotechnical Society. Report by the ad hoc committee for investigating the causes and safety of the dip-slope failure at the 3.1-kilometer mark of Freeway No. 3 sponsored by the Ministry of Transportation and Communications [Internet]. 2011 (in Chinese).

[7] Control Yuan. Public announcement of control yuan on corrective measures of the negligence of the ministry of transportation and communications [Internet]. 2011 . Available from: ge/message_1.asp&ctNode=903&msg_id=3429http://www.cy.gov.tw/sp.asp?xdURL=./di/Message/message_1.asp&ctNode=903&msg_id=3429 [Accessed: Feb. 2016] (in Chinese).

[8] Taiwan Construction Research Institute. Final Report on Technical Services of Slope Engineering of National Highways, sponsored by the Ministry of Transportation and Communications [Internet]. 2014 (in Chinese).

[9] Yang S., Ceylan H., Gopalakrishnan K., Kim S. Smart airport pavement instrumentation and health monitoring. In: 2014 FAA Worldwide Airport Technology Transfer Conference; August 5-7; Galloway, New Jersey, U.S. 2014.

[10] Shih C.H., Yu C.K., Chou C.M., Tsai J.S., Lin J.C., Lin Y.S.. A Study on Unmanned Aerial Vehicles Applied to Disaster Prevention and Rescue: A Case of Landslide on Freeway Side. Journal of Cadastral Survey. 2010; 29(3):17-36 (in Chinese).

[11] Shih T.Y., Teo T.A. On the geometric stability of light weight UAS camera. In: Proceedings of ACRS 2013; Oct. 20-24; Bali, Indonesia. Asia Society of Remote Sensing; 2013.

[12] Zhou G., Wu J. Unmanned aerial vehicle (UAV) data flow processing for natural disaster response. In: American Society of Photogrammetry and Remote Sensing; May 1-5; Reno, Nevada. ASPRS; 2006.

[13] Bendea H.F., Chiabrando F., Tonolo F.G., Marenchino D. Mapping of archaeological areas using a low-cost UAV – the Augusta Bagiennorum Test site. In: XXI International CIPA Symposium; Athens, Greece. 2007. pp. 01–06.

[14] Eisenbeiss H., Lambers K., Sauerbier M., Zhang L. Photogrammetric documentation of an archaeological site (Palpa, Peru) using an autonomous model helicopter. In: Proceedings of the XXth CIPA International Symposium; Sep. 26-Oct. 1; Torino, Italy. International Archives of Photogrammetry, Remote Sensing and Spatial Information Sciences: 2005. pp. 238–243.

[15] Eisenbeiss H., Zhang L. Comparison of DSMs generated from mini UAV imagery and terrestrial laser scanner in a cultural heritage application. In: ISPRS Commission V Symposium; Sep. 25-27; Dresden, Germany. International Archives of Photogrammetry, Remote Sensing and Spatial Information Sciences, 2006. pp. 90–96.

[16] Horcher A., Visser R.J.M. Unmanned aerial vehicles: applications for natural resource management and monitoring. In: Council on Forest Engineering Annual Meeting; April 28–30; Hot Springs (AR), USA. 2004.

[17] Li S. The use of low-altitude unmanned aerial vehicle system in the measurement of large-scale topographic maps — Take the 1:2000 mapping of PuWen as an example. In: 11th South EastAsian Survey Congress and 13th International Surveyors' Congress Innovation towards Sustainability; June 22-24; Malaysia. 2011.

[18] Lin Z.J. UAV for mapping low-altitude photogrammetric survey. In: International Archives of the Photogrammetry, Remote Sensing and Spatial Information Sciences; 37(B1), pp. 1183–1186. 2008.

[19] Tahar K.N., Ahmad A., Wan M., Akib W.A.A., Udin W.S. Unmanned aerial vehicle technology for large scale mapping. In: ISG (International Symposium and Exhibition on Geoinformation) & ISPRS(International Society for Photogrammetry and Remote Sensing); ShahAlam, Malaysia. 2011.

[20] Chang K.T., Liou Y.A., Hwang J.T., Lee G.W., Liu J.K. Using a UAV for collecting high resolution geospatial data for golf course maintenance. In: International Symposium of Remote Sensing (ISRS 2014); Korea, 2014.

[21] Ministry of the Interior. Working Manual of 1/1000 Topographic Mapping via Digital Photogrammetry [Internet]. 2010. Available from: http://www.land.moi.gov.tw/chhtml/index.asp [Accessed: May 2015].

[22] sUAS News. sUAS News [Internet]. Available from: http://www.suasnews.com/ [Accessed: May, 2016].

[23] Agisoft L.L.C. Agisoft PhotoScan User Manual: Professional Edition Version 1.1 [Internet]. 2014. Available from: http://www.agisoft.com/support/tutorials/beginner-level/ [Accessed: June 2015].

[24] Fraser C. S. Digital camera self-calibration. ISPRS Journal of Photogrammetry and Remote Sensing. 1997; 52:149–159.

[25] Shi J.H., Yu K.C., Tsai C.H., Lin C.C., Liu C.L., Lin Y.S. National survey and mapping operation implementing UAV. Internal Research Report, National Land Surveying and Mapping Center, No. 099-301000100G, 2003 (in Chinese).

Chapter 5

Rainfall and Landslide Correlation Analysis and Prediction of Future Rainfall Base on Climate Change

Moung-Jin Lee

Additional information is available at the end of the chapter

http://dx.doi.org/10.5772/64694

Abstract

The aim of this study is to analyze the quantitative relationship between the volume of rainfall and landslide occurrence in South Korea. To predict future rainfall, a future climate scenario was developed by downscaling the regional climate model (RCM) from the global climate model (GCM) based on the Intergovernmental Panel on Climate Change (IPCC) A1B scenario. In this study, for a quantitative analysis of correlation between rainfall and landslides occurrence, data on rainfall and landslides in Korea in the 2000s was analyzed using the correlation between the occurrence of landslides and rainfall volume (daily and accumulated) and the maximum hourly intensity of rainfall. Daily rainfalls exceeding 164.5 mm is categorized as high risk for landslide. A rainfall that continued for 3 days was found to affect the occurrence of landslide in Korea in the 2000s more than any other number of days during which rainfall lasted. The research area for the future climate change scenarios (A1B) covers the entire area of South Korea. Annual average rainfall had increased by 271.23 mm during 1971–2100. The development of downscaling method using GIS and verification with observed data could reduce the uncertainty of future climate change projection.

Keywords: correlation analysis, rainfall, landslide, climate scenario, statistical downscale, verification

1. Introduction

A landslide occurs when part of a slope suddenly collapses because of rapid changes in nature such as a torrential downpour, a typhoon, or an earthquake. In South Korea, the rainy season, extending from June to September, coincides with the period when landslides occur. Casualties and property damage have been on the increase, and recently, there is frequent localized heavy rain of several hundred millimeters.

The Fourth Report of Climate Change Assessment, published by [1], predicted that climate change would persist for several more centuries due to the greenhouse gases that have already been discharged, even though they could be mitigated by efforts such as a reduction in greenhouse gas emissions. The abnormal climate and localized torrential rain caused by climate change may lead to higher landslide rates. An analysis of the causes of landslides, with a multilateral, holistic view, is needed, given this expected increase in landslides due to climate change. Damage from landslides can be minimized by assessing landslide vulnerability [2].

In order to prevent and reduce landslide casualties and property damage caused by climate change, it is necessary to develop scientific methods for landslide analysis related to future climate change. It is important to predict landslide occurrence caused by future changes in rainfall through an understanding of the correlation between past and future rainfall and landslides and to develop methodologies to quantitatively predict changes in rainfall due to climate change. In addition, necessary is advancement in methods to identify landslide locations and a methodology to analyze and verify the relationship between landslides and climate change [3].

Landslide occurrence factors can be divided into internal and external factors. Internal factors include natural ones such as geological structure, topography, soil quality, and forest, while external factors include natural ones, such as rainfall, erosion of rivers and shores, and earthquakes, as well as artificial ones such as cut-embankment, logging, estate development, and quarrying. Landslides readily occur when a slope with internally adverse factors is subject to adverse external factors. In South Korea, landslides are intensified between June and August with localized torrential rain, a period when landslides occur frequently due to concentrated torrential rainfall. Reference [4] researched types and frequency of landslides by intensified torrential rain, centered on the urban areas of Avigliano. Reference [5] studied the frequency of landslides in Ethiopia by analyzing the relation between rainfall and topography, including landscape and forest distribution. Reference [6] analyzed landslide susceptibility based on the distribution of land moisture.

As mentioned above, studies on climate change and landslides' cause of occurrence have been individually conducted. Analysis of landslide occurrence due to future climate change, by relating the two, is still in its early stages, and awaiting global recognition. This study is one of the first attempts to directly correlate analysis of rainfall and landslide occurrence. Future landslide frequency would increase given the consistent effects of climate changes and with studies into the analysis of prediction of future rainfall based on climate scenario, such as the IPCC A1B scenario.

In this study, relations between landslide and rainfall, domestically from 1991 to 2010, are analyzed, and by linking these data to future climate changes, rainfall pattern is analyzed.

http://dx.doi.org/10.5772/64694

2. Methodology and contents

2.1. Analysis of relation between landslide and rainfall in the 2000s

A database for analyzing the correlation between landslides and rainfall in South Korea in the 1990s and 2000s was assembled. The period of collection is 19 years, from July 1991 to December 2010, the data being for landslides nationwide in South Korea. The references for locations and dates of landslides included newspaper articles, national and local media broadcasts, and reports by the Korea Institute of Geoscience & Mineral Resources (KIGAM) and the Korea Institute of Construction Technology (KICT). There is a great deal of data for landslides in the 2000s from various sources, but data for landslides in the 1990s, sourced from relevant academic articles and national media, are limited. This is why this section focuses on the relation between landslides and rainfall in the 2000s, data from the 1990s being included only for reference.

Locations and dates of landslides were extracted from the collected data. A climate database was assembled using observational data from the Automatic Weather Station (AWS) for the area closest to each landslide over the preceding 5 days. Centered on the date of the landslide, the daily rainfall data from 5 days before to 5 days after the landslide was organized. The total organized data comprised 186 landslide locations and daily rainfall. The 186 landslide spots were not averaged out to be the spot of the landslide but indicate points where landslides occurred *en masse*. The analysis of the relation between landslides and rainfall is conducted by two means. First, rainfall data was collected for the same day of the year as each landslide in order to analyze the correlation between rainfall on the day of landslides and that in the 2000s. Second, daily rainfall was added to each following day in order to analyze the relation between landslides and cumulative rainfall. Additionally, in order to identify the effect of daily rain intensity on landslide occurrence, cumulative rainfall for 1, 3, and 5 days were analyzed.

2.2. Prediction of future rainfall change based on climate scenarios

Abnormal weather and concentrated torrential rainfall are becoming more frequent around the world, with concentrated torrential rainfall predicted to occur consistently with climate change [7]. The prediction of future rainfall changes, reflected in climate change, can be divided into predictions using rainfall probability and predictions using future climate change scenarios. A prediction using rainfall probability is for calculating the frequency of occurrence of the same rainfall by inputting past climate data—a probable statistic method based on past data. It can be used with past extreme climate events such as concentrated torrential rainfall and typhoons and can predict to a certain temperature degree. However, it requires processing of data from the same period in adjacent observation spots with the same method in order to do the interpolation because the calculated results are depicted as points representing the observation spots.

Predictions using future climate change scenarios are conducted after the development of the regional climate model (RCM) by individual research centers of the relevant nation from the global climate model (GCM) presented by the Data Distribution Center (DDC) of the IPCC.

The GCM is to predict the future by using the atmospheric general circulation model (AGCM) and humanistic, economic models, predicting temperature increases and changes in rainfall by the prediction of CO_2 discharge. The GCM has several problems: first, future predictions are dependent on the results of the prediction model; second, its spatial resolution is between 200 and 400 km and it is not appropriate for nationwide studies such as the ones covering South Korea; and third, it cannot predict extreme climate events such as concentrated torrential rainfall and typhoons. In order to avoid these problems of the GCM, a study to develop the RCM with its spatial resolution at the regional scale has been conducted. The RCM is to be built on the basis of the GCM by connection with the GCM and humanistic, economic status at regional levels. RCM can be improved to a spatial resolution of 20–25 km, but it is still dependent on the results of the prediction model of the future climate change, and it is hard to predict changes in extreme climate events. This study, in order to improve its spatial resolution, improved the existing spatial interpolation by applying the temperature and rainfall lapse rate to conduct the specification of the future climate change scenario at scales covering South Korea.

2.3. Spatial statistical downscaling of climate change scenario

As with climate change scenarios, the spatial data used in climate change studies should be connected continuously in space. However, the climate data observed in the past and those for predicting the future consist of representative values in the form of points, and for areas without such representative values, the values should be inferred by spatial statistic interpolation. Interpolation indicates a method by which continuous spatial distribution data is built from the inference of values of nonobserved and adjacent spots from the values of observed spots.

Spatial interpolation has various methods of different characteristics. This study selected Co-Kriging, a geostatistics technique in which eigenvalues for the relevant spots are predicted by the linear combination of already-known adjacent values. In this technique, the adjacent, actually measured values are linearly combined for interpolation, and values are estimated by using such statistical methods. Co-Kriging estimates values by the statistical analysis of many adjacent measured values, indicating that it reflects correlative intensity among the adjacent measured values as well as the distance to actually measured values. Co-Kriging is advantageous for identifying overall trends. The applied Co-Kriging equation is as shown in Eq. (1):

$$Z^*(u) - m(u) = \sum_{\alpha}^{n(u)} \lambda_a [Z(u_a) - m(u_a)] \tag{1}$$

where u, u_a are the locations estimated and locations of known data, $Z(u)$, $m(u)$ are the estimation from adjacent data used, $m(u)$, $m(u_a)$ are the estimated values of $Z(u)$ and $Z(u_a)$, and $\lambda_a(u)$ is the Co-Kriging weighted value (weight).

As mentioned, Co-Kriging is used in this study because it is advantageous for identifying trends across a wide area. To calculate estimated values on nonobserved spots, the effects of observed values of the relevant observed spot are more reflected as the relationship of the adjacent observed spots and the linearity becomes closer. In other words, the types of linearity between nonobserved spots and adjacent observed spots are inferred and reflected as inversely proportional to distance and observed values.

Among the topographical factors, altitude has the greatest effect on changes in rainfall and temperature, as climate elements. Temperature, of the climate factors, is particularly affected by altitude, decreasing as altitude increases under the troposphere [8].

When an interpolation of the general geographic information system is applied, accurate estimation is difficult because of the severe skewness of spatial dependence by other factors. In this study, temperature and rainfall serve as variables, and altitude is a factor damaging spatial dependence. In order to overcome this problem, a form of Co-Kriging with which altitude data can be directly considered is selected over general Co-Kriging. Co-Kriging is a method used to interpolate data in the process of spatial estimation. Temperature and rainfall data has a linear correlation to altitude data [8], and thus altitude data can present additional information to reduce inferred measured values in estimating values for temperature and rainfall in nonobserved spots.

When the data used in all the study areas, including digital elevation data, is used with Co-Kriging, it is known to cause uncertainty in getting a Co-Kriging system weight matrix because the correlativity of adjacent altitude data is larger than that of sample values of temperature or rainfall [9]. In addition, altitude data, existing at or adjacent to estimated locations, may conceal the effects of distant altitude data. In order to reduce such errors, this study used collocated Co-Kriging in which additional data can be used globally in the existing Co-Kriging, limited only by additional data values at estimated locations in Eq. (2):

$$Z_{CK}^{*}(u)=\sum_{a=1}^{n(u)}\lambda_{a}^{ck}(u)z(u_{a})+\lambda^{ck}(u)\left[y(u)-m_{Y}+m_{z}\right] \tag{2}$$

where $y(u)$: is the altitude value at location u that is not sampled and $m_y + m_z$ is the average value of altitude data and data of temperature or rainfall.

For the altitude data in this study, a digital map (1:25,000) is used as primary data, and the spatial resolution is transformed to digital elevation model (DEM) data (30 m). The locations of the nation's 75 weather stations are plotted on a map, and the land section of the land cover map (1:25,000) is used to build the data for the coastline of South Korea. Because the observed values from the 75 weather observation spots between 1971 and 2000 differ in altitude above sea level and in the heights above surface level of thermometers and rain gauges, a modification equation for the temperature lapse rate was presented by [10]. The temperature lapse rate was based on altitude (Eqs. (3) and (4)). As for rainfall data, the rainfall lapse rate is considered on the basis of altitude [11] (Eqs. (5) and (6)). Eq. (5) shows that the rainfall value should be increased by 74% per 1 km for October-April (Cold Season). For the period May-September (Warm Season), Eq. (6) reflects a 46% reduced rainfall value:

$$|\tau| = 0.00688 + 0.0015\cos 0.0172(i-60)\text{Average Temperature Lapse Rate} \quad (3)$$

where |τ| is the absolute value of air temperature lapse rate based on annual dates, and i is the annual date (1/1 day = 1, 12/31 day = 365):

$$\mathrm{T} = T_i + \text{Elevation}(m) \times |\tau| \quad (4)$$

where T : is the temperature corrected by the air temperature lapse rate, and T_i: is the daily temperature (air-temperature lapse rate corrected before the temperature):

$$R = R_i \times (1.74)^{\frac{E}{1000}} \quad (5)$$

$$R = R_i \times (0.46)^{\frac{E}{1000}} \quad (6)$$

where R represents rainfall and E represents elevation (m).

3. Result

3.1. Relation between landslide and rainfall of the day

Figure 1 shows representative types of landslides and rainfall. The X-axis averages periods of rain and the Y-axis averages periods of occurrence per year as frequent rainfall. Rainfall in the diagram is calculated from the AWS's rainfall data within the areas of landslides.

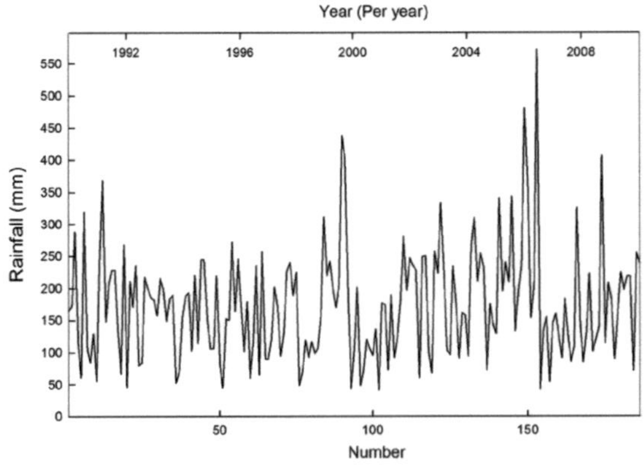

Figure 1. 1-day maximum rainfall of the study area in the 2000s.

http://dx.doi.org/10.5772/64694

Regarding the characteristics of rainfall on days related to landslides in the 2000s, landslide dates are clustered around July 14 and 16, July 22 and 23, August 25 and 26, and September 13 and 16. Examining average rainfall at the time of landslide occurrence, landslides seem to usually occur when rainfall is around 170 mm. Landslides occurring when rainfall was less than 100 mm are considered to result from temporary construction or high cumulative rainfall.

3.2. Relation between landslide and cumulative rainfall

A comparison of cumulative rainfall before landslides with rainfall on the day of landslides was conducted in order to analyze the relation between rainfall on the day of a landslide and cumulative rainfall as a possible cause of landslides. Cumulative rainfall before landslides (1, 3, and 5 days) and rainfall on the day of landslides are compared per year. The horizontal (*X*) axis indicates rainfall on the day of the landslide, while the vertical (*Y*) axis indicates the amount of accumulated rainfall over 1, 3, and 5 days prior to the landslide. **Figures 1–4** show that landslides are caused by rainfall when a region over the 45° central border line is spotted, while they are considered to be caused by cumulative rainfall when a region under the line is spotted. In an analysis of rainfall on the day of a landslide and the cumulative rainfall of 1 day before the landside for the 2000s, landslide spots over and under the 45° central border line are clearly separated. There are 162 spots over and 24 spots under the central border lines (**Figure 2** and **Table 1**). **Figure 2** and **Table 1** include rainfall on the day of the landslide and cumulative rainfall for 3 days before the landslide, where the spots are distributed over and under the 45° central border line, indicating that landslides are equally caused (spots over the border: 86; spots under the border: 80) by rainfall on the day of the landslide. The cumulative rainfall data for 3 days prior is shown in **Figure 3** and **Table 1**. Rainfall on the day of the landslide and cumulative rainfall for 5 days prior are shown in **Figure 4** and **Table 1**. Landslides are spotted more on the area under rather than over the 45° central border line (spots over the border: 56; spots under the border: 130), indicating that landslides are caused by 5-day cumulative rainfall rather than by rainfall on the day of occurrence. Note that 3-day cumulative rainfalls, therefore, may be more indicative of whether landslides are caused by rainfall on the day of occurrence or by cumulative rainfall. Many landslides in the 2000s are closely related

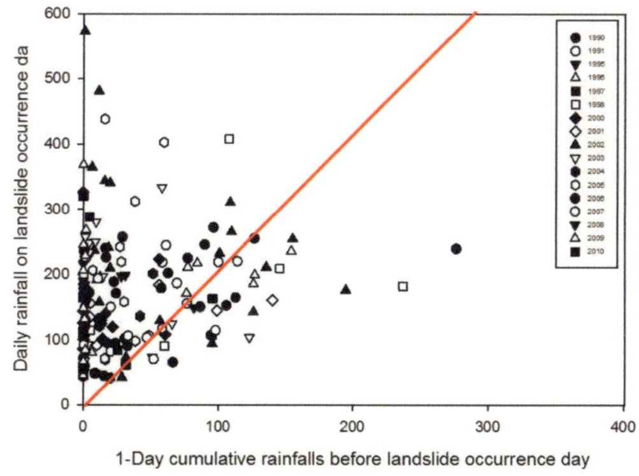

Figure 2. 1-day cumulative rainfall before landslide occurrence day.

to 3-day cumulative rainfall, and thus landslides in the 2000s may be affected more by cumulative rainfall than by rainfall on the day of occurrence. As a result, 1-day rainfall and 3-day cumulative rainfall, rather than cumulative rainfall on other days, had a higher correlation with landslide occurrence.

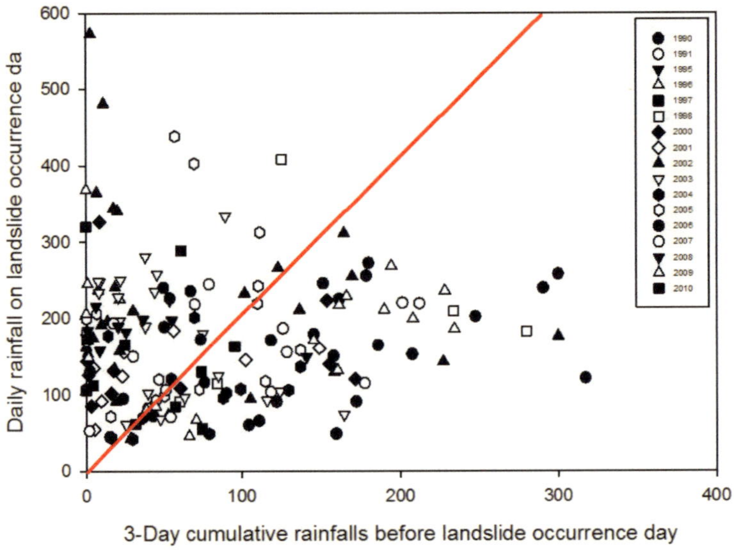

Figure 3. 3-day cumulative rainfall before landslide occurrence day.

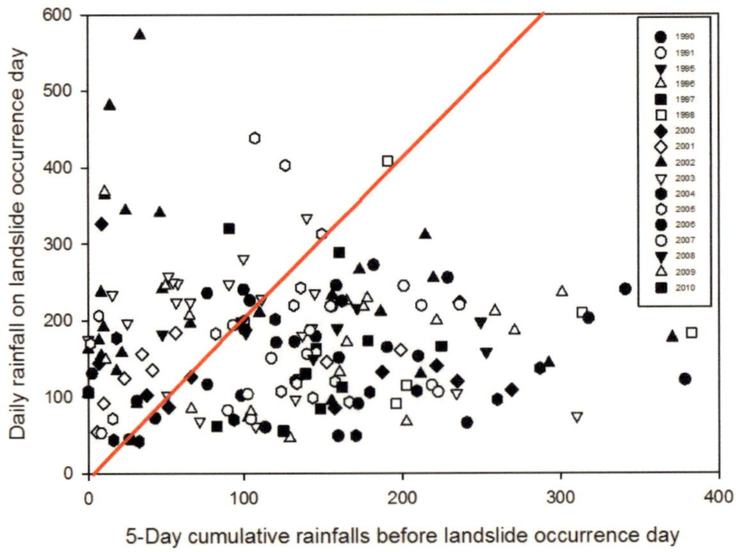

Figure 4. 5-day cumulative rainfall before landslide occurrence day.

http://dx.doi.org/10.5772/64694

Classification	Over	Under
Daily rainfall and 1-day cumulative rainfalls before failure data (**Figure 2**)	162	24
Daily rainfall and 3-day cumulative rainfalls before failure data (**Figure 3**)	86	80
Daily rainfall and 5-day cumulative rainfalls before failure data (**Figure 4**)	56	130

Table 1. Summary of landslide spot counts over and under the 45° central border line (number of spots).

3.3. Scenario for future climate changes in South Korea

The period for the study results built from spatial statistical downscaling of the KMA-RCM is 1971–2100, and average temperature and rainfall per month are produced. Temperature and rainfall are selected from the climate change models because they are the two factors that can express future changes in climate fragmentally and representatively, with higher practical utility (**Figure 5**).

Rainfall was analyzed by accumulation of the monthly average, decreasing from 947.38 mm in Year 1971 to 886.02 mm in Year 2000, a drop of 61.35 mm. Results of future climate change data processing showed that the rainfall estimate increased from 1002.12 mm in Year 2001 to 1218.60 mm in Year 2100, or by 216.48 mm. The annual average rainfall is showed by 271.23 mm from Year 1971 to Year 2100 (**Figure 6**). **Figure 6**, where the X-axis is the year and the Y-axis is average rainfall, respectively, shows a linear increase.

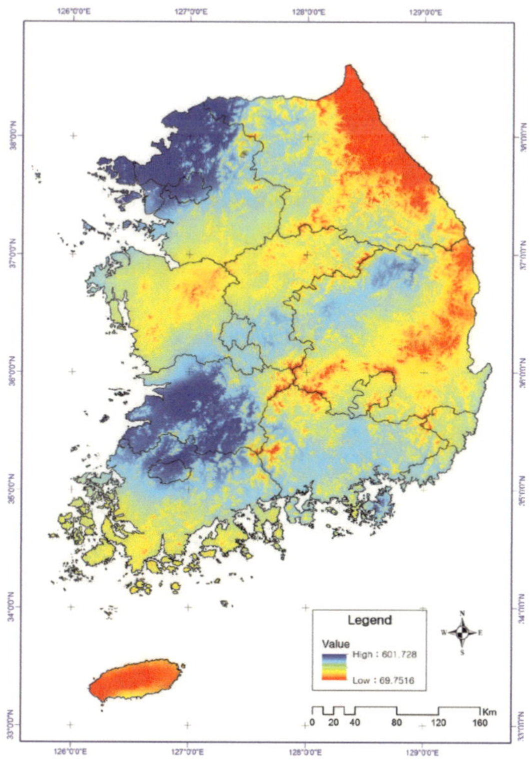

Figure 5. The average rainfall in July 2071 (based on KMA-RCM).

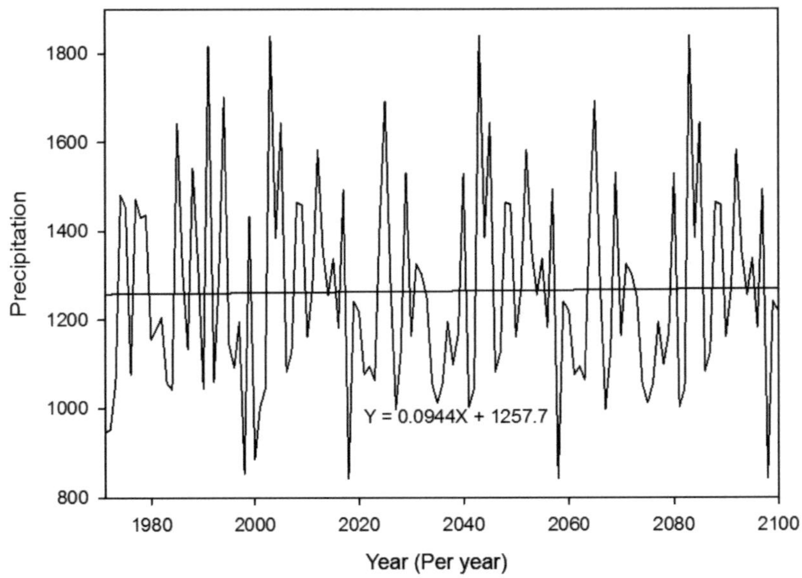

Figure 6. 1971–2100 annual rainfall change in South Korea.

Regarding the annual cumulative rainfall for this 130-year period, 1839.36 mm (2003, 2043, and 2083 year) is the highest, and 847.37 mm (2018, 2058, and 2098 year) is the lowest, a difference of 991.99 mm, while 1817.03 mm (Year 1991) is the second highest, and 853.5 mm (Year 1998) is the second lowest. When years with the same rainfall distribution were organized, 71 are separated. In the distribution maps, built from past climate data, single rainfall is shown as 1817.03 mm and 1701.14 mm. When downscaling based on the KMA-RCM is conducted, the changes in cumulative rainfall as a future change in climate show the repetition of the same distribution with an interval of 40 years, with an overall increase in rainfall [12].

3.4. 1971–2100 annual rainfall change in South Korea: verification of climate change scenario result

In order to reduce uncertainty occurred in the process of spatial statistical downscaling of future climate change scenarios, the analysis of the correlation between this research's results and actual measurements was carried out. Weather data after Year 2000 was selected as the reference year and 75 AWS, which were closest to 75 ground observation points, and data from January 2001 to August 2010 existed was selected and the analysis of correlation between temperature and rainfall data and data obtained through future climate change scenario prediction from 2001 to 2010 was carried out. The data was analyzed and the coefficient of correlation between future climate change scenario prediction data and actual measurements from AWS was over 0.98 on average in the case of temperature and 0.56 in the case of rainfall. This was lower than that of the temperature, but in consideration of uncertainty in rainfall prediction on the climate change scenario, it was a high correlation.

Area	Correlation coefficient	Area	Correlation coefficient	Area	Correlation coefficient
Gangneung	0.41	Busan	0.63	Jangheung	0.49
Ganghwa	0.70	Sancheong	0.53	Jeongu	0.65
Geochang	0.52	Seosan	0.50	Jeongeup	0.65
Goheung	0.49	Seoul	0.64	Jecheon	0.64
Gwangju	0.46	Sokcho	0.45	Jinju	0.48
Gumi	0.62	Suwon	0.62	Cheonan	0.65
Gunsan	0.61	Suncheon	0.61	Cheorwon	0.68
Geumsan	0.60	Andong	0.63	Chupungnyeong	0.67
Namwon	0.57	Yangpyeong	0.62	Chuncheon	0.67
Namhae	0.40	Yeongdeok	0.45	Chungju	0.58
Daegwallyeong	0.59	Yeongwol	0.60	Taebaek	0.57
Daegu	0.49	Yeongju	0.64	Tongyeong	0.39
Daejeon	0.73	Yeongcheon	0.50	Pohang	0.45
Masan	0.45	Ulsan	0.46	Hapcheon	0.49
Mungyeong	0.70	Uljin	0.35	Haenam	0.49
Baengnyeongdo	0.50	Uiseong	0.60	Hongcheon	0.60
Boyeong	0.59	Icheon	0.57	Heuksando	0.50
Boeun	0.76	Incheon	0.64	Yeosu	0.42
Bonghwa	0.66	Imsil	0.66		
Busan	0.29	Jangsu	0.65		

Table 2. Result of correlation between statistical downscaling of future climate change scenario and AWS in Years 2001–2010 (rainfall).

Note that 75 AWS where rainfall data existed from January 2001 to August 2010, 75 observation points were extracted and summarized in **Table 2**. In the case of correlation on rainfall summarized in **Table 2**, most weather stations showed a significantly high correlation, but in some cases of correlation on rainfall shown in **Table 2**, the range of rainfall variability was significantly high depending on the weather station.

In the analysis of correlation on the amount of rainfall, Ganghwa showed a correlation of 0.70, which was significantly high, but Busan showed a correlation of 0.29, which was lower than that of other areas. It was analyzed that generally the correlation between future climate change scenario prediction data and actual measurements was significantly high, but in the case of rainfall, the variability for each year and season was high. South Korea shows abundant rainfall in summer, and the standard deviation of rainfall in summer varies more than two times between regions, and the rainfall variability for each region is also significantly high. In addition, the range of rainfall variability near the coast is higher than that in the inland area in South Korea, and the east coast has the higher rainfall variability than the west coast [13]. In the result of correlation between the rainfall and future climate change scenario prediction data, the bottom 10%, which had the lowest correlation, showed a correlation of less than 0.40, and the corresponding area included Busan, Uljin, Tongyeong, and Namhae. In order to analyze the correlation of relevant area more precisely, the analysis of correlation for each season was carried out additionally. An area with a low correlation was selected because a seasonal change in an area with a low correlation was more closely related with rainfall change [13]. In the analysis result, it was analyzed that the correlation between three regions except for Uljin was lower during summer and winter, with a high range of variance in the meteorological factors (**Table 3**).

Area	Correlation coefficient	Correlation coefficient (summer and winter)	Correlation coefficient (spring and fall)
Busan	0.29	0.24	0.45
Uljin	0.35	0.30	0.16
Tongyeong	0.39	0.37	0.50
Namhae	0.40	0.39	0.43

Table 3. Result of seasonal correlation coefficient between statistical downscaling of future climate change scenario and AWS in Years 2001–2010 (rainfall).

4. Discussion and conclusion

In this study, the pattern of rainfall generating a landslide in the past was analyzed, and the threshold of landslide occurrence by rainfall was also analyzed. Based on the analysis result, the possibility of landslide occurrence in the future was analyzed by analyzing the rainfall of future climate change scenarios.

When the relation between rainfall during the 2000s and landslides in South Korea is quantitatively analyzed, 1-day rainfall and 3-day cumulative rainfall had higher correlations with landslides. Based on this, the landslide occurrence threshold in the study area is defined to be 202 mm for 1-day rainfalls and 449 mm for 3-day cumulative rainfalls. The results of this analysis of rainfall probability show the ratio of the occurrence threshold consistently increases as the target year increases in the study area, the same tendency is seen in the future climate change scenario. Conclusively, the study area has seen increasing rainfall as time passes, and damage, such as that from the 2006 landslide, may increase gradually in the future.

As the target year increases, the accuracy increases for the 202–449 mm thresholds of rainfall probability and the future climate change scenario. In all of the methods applied, the accuracy is higher for the 449 mm threshold than for 202 mm of rainfall probability. A 3-day cumulative rainfall affects landslide occurrence more than a 1-day rainfall in the relation between rain and landslides.

In addition, a rainfall change from Years 1971 to 2100 was analyzed through the analysis of the climate change scenario based on KMA-RCM in this study. The main result of this study can be summarized as follows.

First, the downscaling technique using rainfall lapse-rate technique was developed by using Co-Kriging among geographic information spatial-interpolation techniques. As a research result, the average rainfall between 1971 and 2100 was drawn. The result showed that the average rainfall increased by 271.23 mm.

Second, the analysis of correlation between the average rainfall between 2001 and 2010 and actual measurements from 75 AWSs from regional-scale climate change scenario was carried out. As a result of analyzing the correlation between the future climate change scenario prediction data and actual measurements during the same period, it was concluded that the correlation on the rainfall was 0.56.

A study to downscale KMA-RCM with a spatial resolution of 27 km into the climate change scenario with a spatial resolution of 1 km, which could express the local level climate change, was carried out. In order to reduce the uncertainty in the climate change scenario occurred during this process, the correlation between actual measurements after the reference year entered on the future climate change scenario and research results were analyzed. The result verified that the significance of downscaling of results through KMA-RCM is high, and the average climate pattern (30 years in the past) and the weather pattern from Years 2001 to 2010 were similar. In the case of rainfall, temperature, humidity, and local characteristics (e.g., topography and characteristics of ground surface, etc.) have interacted in the rainfall process complexly, showing a nonlinear relationship [13], and the result of this study was also affected in the same way.

The results of this analysis of correlation between rainfall and landslide show the cumulative rainfall consistently increases in South Korea, the same tendency is also seen in the future climate change scenario.

A result of correlation between rainfall and landslide and the future rainfall change also shows future rain events for quantitative analysis of climate change in South Korea. Changes in rainfall in South Korea are shown to be larger.

The occurrence of landslides is directly caused by intensive rainfall. If there is a change in rainfall, it will lead to a change in the occurrence of landslides.

However, this study is unable to reflect extreme conditions according to climate change, and, even though, the correlation between the result of this study and actual measurements was significantly high, it is necessary to improve the methodology to future climate change scenario prediction continuously. It is necessary to supplement methodologies regarding extreme rainfall and extreme climate events in order to reduce the uncertainty in the future climate change scenario in future studies.

Acknowledgements

This research was supported by Korea Environment Institute funded by the Basic Science Research Program through the National Research Foundation of Korea (NRF) funded by the Ministry of Education (NRF- 502014R1A1A1002704).

Author details

Moung-Jin Lee

Address all correspondence to: leemj@kei.re.kr

Korea Adaptation Center for Climate Change(KACCC), Korea Environment Institute(KEI), Sejong-si, Korea

References

[1] IPCC. Climate Change 2007—IPCC 4th Assessment Report, Working Group 1—The Physical Science basics. IPCC; 2007. p. 996

[2] Creutin J D, Delrieu G, Lebel T. Rain measurement by raingage-radar combination: a geostatistical approach. Journal of Atmospheric and Oceanic Technologies. 1998. 5: 102–115. DOI: http://dx.doi.org/10.1175/1520-0426(1988)005<0102:RMBRRC>2.0.CO;2

[3] Daly C, Helmer E H, Maya Quinones. Mapping the climate of Puerto Rico, Vieques and Culebra. International Journal of Climatology. 2003. 23: 1359–1381. DOI: 10.1002/joc.937

[4] Borge M. Accuracy of radar rainfall estimates for streamflow simulation. Journal of Hydrology. 2002. 267: 26–39. DOI: 10.1016/S0022-1694(02)00137-3

[5] Temesgen B, Mohammed M U, Korme T. Natural hazard assessment using GIS and remote sensing methods, with particular reference to the landslides in the Wondogenet Area, Ethiopia. Physics and Chemistry of the Earth, Part C: Solar, Terrestrial & Planetary Science. 2001. 26: 665–675. DOI: 10.1016/S1464-1917(01)00065-4

[6] Regmi N R, Giardino J R, Vitek J D. Modeling susceptibility to landslides using the weight of evidence approach: Western Colorado, USA. Geomorphology. 2010. 115: 172–187. DOI: 10.1016/j.geomorph.2009.10.002

[7] Daly C. Guidelines for assessing the suitability of spatial climate data sets. International Journal of Climatology. 2006. 26: 707–721. DOI: 10.1002/joc.1322

[8] Benestad R E. Empirically downscaled multimodel ensemble temperature and precipitation scenarios for Norway. Journal of Climate. 2002. 15: 3008–3027. DOI: http://dx.doi.org/10.1175/1520-0442(2002)015<3008:EDMETA>2.0.CO;2

[9] Hewitson B C, Crane R G. Consensus between GCM climate change projections with empirical downscaling: precipitation downscaling over South Africa. International Journal of Climatology. 2006. 26: 1315–1337. DOI: 10.1002/joc.1314

[10] Comriea A C, Broylesb B. Variability and spatial modeling of fine-scale precipitation data for the Sonoran Desert of south-west Arizona. Journal of Arid Environments. 2002. 50(4): 573–592. DOI: 10.1006/jare.2001.0866

[11] Smith C D, The relationship between monthly precipitation and elevation in the Alberta Foothills during the Foothills orographic precipitation experiment. Cold Region Atmospheric and Hydrologic Studies; Chapter 10. The Mackenzie GEWEX Experience. Springer Berlin Heidelberg, 2008. pp. 167–185. DOI: 10.1007/978-3-540-73936-4_10

[12] IPCC TGICA. General Guidelines on the use of Scenario Data for Climate Impact and Adaptation Assessment. Version 2. IPCC; 2007. pp. 120–122

[13] Lee M J, Park I, Won J S, Lee S. Landslide hazard mapping considering rainfall probability in Inje, Korea. Geomatics, Natural Hazards and Risk. 2016. 7(1): 424–446 DOI: 10.1080/19475705.2014.931307

Section 3

Risk Assessment of Manmade Geohazards

Chapter 6

The Long Shadow of Human-Generated Geohazards: Risks and Crises

Franco Oboni and Cesar Oboni

Additional information is available at the end of the chapter

http://dx.doi.org/10.5772/66066

Abstract

The purpose of this chapter is to focus attention on the "damage and risk" side of the geohazard (GHZ) phenomena rather than on their generating processes. Damage evaluations are indeed often neglected and oversimplified in predictive studies. As a result, risks are poorly understood and often considered as the mere expression of the probability or likelihood of an adverse event. In this chapter, we will use numerous real-life examples and will discuss among other subjects: technical glossary of risk, damages, crises, multidimensional consequences analysis, and definition of risk tolerance. This chapter also focuses on ethical (geo-ethical) issues linked to GHZs caused by human activities and their mitigation decisions and possible unintended consequences. The discussion includes the sometimes excessive and sometimes lacking (blindness) perception of risks by the public, corporate, and public officers. The root cause of some odd human behaviors when facing risks (biases) like the survivor bias is discussed. GHZs cast a long and often misunderstood shadow on human activities, development, and survival. By understanding how to model consequences and better evaluating risks and crises, we will be able to alleviate human and environmental suffering and foster sustainable development.

Keywords: Risk, Crises, Social, Interdependencies, Consequences, Glossary, tolerance, geoethics, anthropocene

1. Introduction

Geohazards (GHZs) are defined as geological (geotechnical, hydrogeological) states that may lead to widespread damage or risk. They are geological and environmental conditions involving long-term or short-term geological processes. Humans are also altering the planet with actions, traceable back to the Neolithic age agriculture. We can call these interactions between the geosphere and humanity Anthropogenic (ANPgenic) global changes.

As we can safely assume we are in the Anthropocene (ANPcene), it is also safe to assume that many of the processes we observe nowadays are man-made or man-altered. It is therefore obvious that risks and crises, hazard and risk perception and related decision-making, loaded with their social aspects, have to be studied in a holistic way, integrating humanistic and social aspects to rational, technical, and scientific approaches.

Human-generated, i.e., man-made or man-altered, GHZs and natural-occurring GHZs lists can be, in first analysis, assumed to be identical. That applies even to seismicity as lately exposed by research on fracking and other oil-extraction techniques. Exceptions are special hazards like the spread of unexploded landmines contamination via erosion and flooding processes, the spread of heavy metals or other contaminants via dumps leaching or mining tailings, dam failures, etc.

This chapter focuses attention on the "damage and risk side" of the phenomena rather than on its generating processes or on the full risk assessment/management (RA, RM) process: GHZ processes may hit targets T with a probability p and generate damages (losses) or consequences, C. Thus, the generated risk on T is p*C.

Damage (consequence) evaluations are often cursorily discussed by geoscientists, and obviously so, as those themes are not within their direct scope of knowledge. Scientists are of course more interested in studying the science behind processes rather than their potentially grim outcomes. Thus, damage evaluations are often oversimplified by predictive studies, and risks are poorly understood.

This chapter first reviews the technical glossary of risk, damages, and crises using real-life examples. Various concepts including ways to evaluate probabilities, frequencies, and their relationship are then approached. Further sections bear on multidimensional consequences analysis and risk tolerance. The discussion includes accident consequences and risk perception; identical consequences generating different behavior; and third-party hazards possibly requiring strategic shifts if common mitigations are impossible.

Ethical (geo-ethical) issues linked to GHZs caused by human activities and their mitigation decisions are discussed with special focus on possible unintended consequences and the sometimes excessive, sometimes lacking (blindness) perception of risks by the public, corporate, and public officers. The root cause of some odd human behaviors when facing risks (biases), like the survivor bias, is then discussed.

We present below a navigation help for the reader, which: (a) summarizes some very general rules to consider when building an RA model [1, 2] while (b) displaying what is included in this chapter and where to find it. The items marked with "literature" are an invitation for the reader to search the vast body of specialized literature on specific RM themes that could not enter in the chapter for obvious space limitations.

Any RM model should satisfy all five roles of system science, namely:

http://dx.doi.org/10.5772/66066

- describe the physical world **[Glossary (Section 2.1); System Description (Section 2.2)];**
- portray the results of interactions among a few of its components **[Consequence Analysis (Section 3.2)]**;
- propose a generic design **[Mitigations:(Section 4.5.2 and Literature)];**
- be a constituent of "science of complexity" as it enlarges the domain of demonstrable results in the service of humanity [**Interdependencies (Section 4.4)**]; and
- be actionable, as it has linguistic clarity and a model that suggests a clear direction of actions essential to resolve emergencies **(Decision-Making: Literature).**

Any RA/RM model should also address some of the Complexity Laws [2], like:

- not require humans to process more than three components at a time (triadic constraint) **[e.g., Probability (Section 3.1), Consequence, Risk and Perception (Section 4.)];**
- render a parsimonious description of any emergency **[Unified Scale of Consequence (Section 3.2)];**
- address the challenge of vertical incoherence as it can show the right aggregated level to decision-makers at different organizational levels **[Interdependencies (Section 4.4), Communication, Information Dashboards (Section 2.3.2) and Literature];**
- consider all relevant factors of emergencies in a balanced fashion **(Rational Risk Prioritization: Literature)** [3].

2. Need for technical glossary, clear semantics of risk, damages and crises

Over two decades of day-to-day work in the space of GHZs, risk assessment (RA) and risk & crisis management (RM, CM) have shown that many problems, oftentimes leading to ill-conditioned decision-making, squandering of private and public resources (money), misaligned expectations, boycotts, protests, and even turmoil arise from: (a) unclear glossary, (b) poor semantics, (c) poor (or absent) definition of the system(s) to be assessed, and (d) use of misleading methods or poor understanding of extant methodologies.

2.1. Glossary and semantic of risk statements

In our capacity of third-party reviewers, we encounter many "classic" missteps in industrial or governmental agencies' RAs around the world [4]. In this chapter, we expose a number of these missteps following the logic described below.

Misstep,
Erroneous statement,
Rule (numbered),
CORRECT example.

We will start with a misstep stemming from the use of confused technical glossary.

Misstep example: We have heard people talking about "risk" as a synonym for probability or hazard. This becomes extremely confusing when modeling a system and discussing what are manageable/unmanageable risk and how to address them. Hazards, in short, are anything that can go wrong. Hazards have a probability of occurring and potential consequences. Risks are hazards' probability*consequence.

Erroneous statement: "The risk of dam A breach is 10^{-4}." It is wrong because 10^{-4} is a probability of occurrence and not a risk. Risk is p*C, as noted in the Introduction section and above.

Rule 1: Always use a well-defined technical glossary throughout the studies and presentations. Do not accept improper or unclear definitions from anyone. Do not try to guess the meaning of other team members.

Quick fix: Always base your assessment on a well-defined glossary, for example, see http://www.riskope.com/knowledge-centre/tool-box/glossary/. There are many others available in literature.

Correct: the risk of dam A breach is 10^{-4}, with a consequence of......casualties, ...environmental damages, and...lost infrastructure, business interruption, etc. As you can see, "talking" risk is not that simple, as despite the simplicity of the governing equation, there are numerous nuances. Shortcuts are deadly as they create confusion.

If one needs to prove the above statement, it is enough to look at the questionnaires that large consulting companies of this world send out to "get the temperature" of RM to managers without specifying the glossary first. The replies they get, often bound-up in nice "yearly reports" on the state of risks in the world, are misleading to say the least!

Due to ubiquitous confusion, modern decision-makers (DMs) are reportedly feeling the need to improve measurement and risks communication in various areas, as stated in a recent international consulting company report. Better evaluation/definition of key risks and enhanced definition of organizational purpose and values are at the top of listed concerns. Decision-makers (DMs) want better communication and hopefully politician will too, if they are not too concerned by electoral pressure.

DMs seem to realize that RAs based on failure modes and effect analysis (FMEA) and other probability impact graphs (PIGs, see **Figure 1**) only clutter their horizon, leaving them struggling, and realize that it is difficult to communicate organizational purpose and values if one does not know the risks to which the company exposes itself or exposes the public. No wonder "modern" DMs are looking to improve measurement and risks communication, and let us hope it is not too late from an ANPcenic point of view.

Naming risks by their consequences (e.g., environmental risks, frequently used to define mishaps that could lead (among others to) environmental consequences) constitute a misleading (for the analyst) and confusing practice for the user/DM.

http://dx.doi.org/10.5772/66066

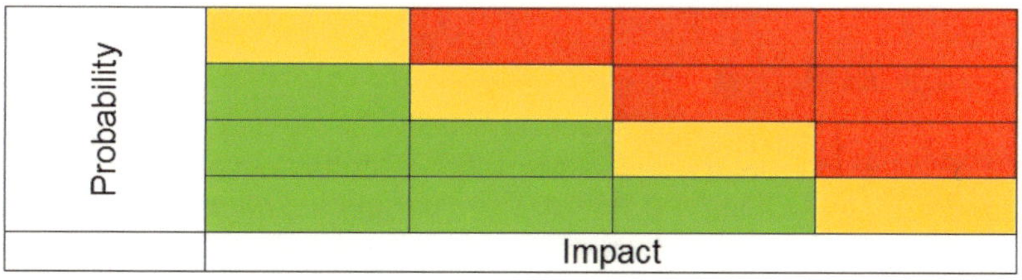

Figure 1. Example of Probability Impact Graph (PIG).

2.2. Defining the Anthropocenic system to be assessed

As stated in Section 1, prehistoric and historical evolution of humankind meant to modify environments to appropriate resources. The timescale of the modification of the environment is briefly reviewed by looking at "the state of the world" thousands of years ago, 1000, 500, and 200 years ago, and then by imagining the requirements in those times, and their implications for the long term. We have selected a mining operation, which would have been built then [5].

2.2.1. A brief review of time

If our mining company had existed:

Forty-five thousand years ago, we would have been mining hematite at Bomvu Ridge in Swaziland. Our mine would have produced insignificant waste and no societal concerns.

Three thousand years ago, we would have been bidding for the rights to mine silver at Laurentia just east of Athens. Slaves would have worked to our command to provide the money to build the Parthenon.

One thousand years ago, we would probably have been asked to sponsor the crusades. Total world population was around 300 M souls. Languages used then have since disappeared.

Five hundred years ago, we could have learned about Mr. Columbus' recent discovery. Total world population had increased to 500 M souls.

Two hundred years ago, we would have been concerned by the Battle of Trafalgar, Napoleon's retreat from Moscow, and the Battle of Waterloo. The world's population had reached 950 M souls.

If we had closed our mining tailings facility 1000, 500, or 200 years ago, would we have expected that the tailings (mining wastes) should still be right there where we dumped them, unattended, not maintained, not monitored? Oh, we were forgetting one thing: had we left a Standard Operating Procedure and Maintenance Manual for "future generations," now the manual would be in a difficult (impossible) to understand language. The documents might have turned to dust or have been heavily damaged. In addition, if we think digital transcriptions of our documents may have saved us, well the solar flare of 1859 (Carrington event) would probably have erased them all if fires, floods, and wars had not done it earlier.

Keep the information above in mind as you will go through the rest of this chapter.

2.2.2. Physical system model

Defining the system to be assessed is the key for a meaningful RA and yet one of the most neglected phases under the common excuse that ANPgenic systems are "too complex." We are not denying complexity, but people tend to cross their arms and "do nothing" because they encumber themselves with clouded preconceived opinions. We all know that ISO and other international and national risk codes stress the fact that the context of the study and the environment in which a system operates has to be described. However, so many times, we have seen project teams and facilitators embarking in FMEAs or other risk-related endeavors without taking the time to rigorously describe the system anatomy and physiology. Although it may seem inappropriate to use medical terms, you will see they are very useful for illustrating purposes.

A brief history of medicine: In ancient times, human health (the system of interest in medical science) was in the hand of shamans and other medicine men (and women). Visibly, the understanding they had of human body was not satisfactory. They needed to understand more. Hence, for example, Leonardo da Vinci started to perform anatomical studies (dissection was prohibited by the Church and the Law in those times) and recorded his acute observations in the famous sketches still displayed in various museums around the world. Those studies delivered a first understanding of human anatomy. A few more centuries of research brought us to be able to detect genetic mutations, hereditary diseases, and much more. Only in the early 1900, thanks to S. Freud, we started treating psycho-pathologies with psychoanalysis and then started understanding the link between physical ailments and psychological troubles.

A brief history of RA methods: This story is faster and shorter than the prior one. Most common-practice tools have military origins as they were used to increase weapons reliability. Industry was still using the so-called insurance gals two decades ago to transfer risk, without any serious evaluations, to insurance companies willing to take a bet on them. Then, a series of GHZ and man-made mishaps, public outcry, and political pressure transformed "risk" in a fashionable buzz-word. RA and RM were nice words to say and common practice percolated down to the minimum common denominator, using FMEA and other methods and models to administer social "placebo." Accidents were still occurring, failures were still considered "unforeseeable," and potential consequences were still looked at cursorily and in a compartmentalized way. No one was carefully describing the system's anatomy and physiology in industry and GHZs RAs. It was the time of open-risk workshops where participants were able to voice concerns and fears, without having dissected the system under consideration, pretty much like we used to do in medicine before understanding anatomy and physiology. Then, large-scale terror acts (September 11, 2001) occurred on U.S. soil and, in 2008, there was a global recession. All of a sudden, new words were coined to describe what we all knew very well already: poorly made RAs do not bring any value to projects and society. The talk was all about systemic risk, nonfunctioning models, Black Swans (BS, see Section 3.1), fragility, complexity, etc. All of those efforts just to hide one simple fact: unless we take the time and effort to properly define our systems, we cannot perform any serious analysis on them! The parallel is striking: if we do not

http://dx.doi.org/10.5772/66066

know the human body anatomy and physiology, any surgery or drug will have a very poor rate of success, or be detrimental. Consider GHZs diseases of our ANPcenic systems and you will be ahead of the game.

Misstep example: Put together a risk register without defining the system's functional analysis, success/failure criteria. Consider self-sufficient engineering systems forgetting their interactions with other systems/subsystems, the environment and the world.

The largest and costliest mistakes are generally made when (poorly) defining the system. You have to understand the context of the study and what constitutes the system you have to assess.

Erroneous statement: Let's perform an RA of Lake B environment, but let's not look at what could happen upstream of the contributing rivers and creeks.

Rule 2: Always perform a functional analysis (as is required, but very seldom performed, when starting a FMEA study). Be sure to take into account cascading failures and inter-systems interdependencies. The definition of the success/failure criteria is fundamental to understand both the hazard and the system.

Quick fix: Determine the limit of the system and the logical connections between the components. Then state why inclusion/rejection decisions are made as part of increasing study's transparency.

Correct: Let's include in the RA of Lake B environment all the contributing rivers and creeks, contamination that comes from aerosols and air pollution, and what could be mobilized (e.g., bottom purge of reservoirs that could have accumulated mercury or dioxins in sediments).

Misstep example: Starting the RM process by brainstorming all possible risks with the crews, without proper preparation, most of the time leads to mislabeling hazards or concerns as "risks" (See Rule 1).

Erroneous statement: Fire is a significant risk in this facility. This statement misses two major points. What fire: a wildfire (like the one Fort McMurray, Canada) or an electrical fire (man-made) or an arson? Where and hitting what?

Rule 3: Always start by identifying hazards using threats-to and threats-from. Perform strong logic checks on your risks definitions.

Quick fix: The hazard identification process is an important step, but not the first one. Only once all the logical connections are established, one can be sure that you have been as methodical and exhaustive as possible.

Correct: There are several hazards potentially causing fires at this facility: natural conditions (including climate change), operational mistakes by subcontractors (actually the lack of instruction of the subcontractor would be the root cause), and criminal/terror activities. The potential targets of each would be.... the consequences of each at every potential target would be …

Working properly with a robust logic leads to a significant increase of the number of records of the Hazard and Risk Register (HRR) (the database that should contain all your RA information based on your physical model). The efficiency of off-the-shelf all-purpose worksheet software is limited, and specific software solutions have to be sought.

Misstep example: If in a HRR a hazard is listed without a "threat-to," it is impossible to assess its consequences. The same hazard (say a rock falling) can lead to the definition of a widely different risk because the consequences may vary in time and location.

Erroneous statement: "Traffic accidents" in this industrial area have a consequence ranging from … to … This statement is wrong because the hazard will have different consequences depending on what it is impinging on and what generates it (construction equipment hits pipeline, snow removal knocks out a data telemetry station, etc.). In many cases, consequences (or the hazard) will be counted again in another scenario.

Rule 4: Check your risk statements (record per record in your hazard and risk register) to avoid double counting.

Quick fix: Always link a hazard to a component of the system. If various hazards can hit the same component, or if the same hazard can hit many components, each one of them has its own line in the hazard register.

Correct: Traffic hazards by fork-lifts on acid pipeline have a consequences ranging from … to …; traffic hazard by subcontractors trucks …, etc.

In Section 4.4, we discuss third-party-induced risks through interdependencies. As we will see, those constitute a special case of threat-from analysis.

A key element for the success of an RA is the definition of the success criteria (i.e., the RA's metric). Not achieving the success criteria means being in a failed state. Clear definition of success/failure constitutes the basis for rational RAs as it allows to understand what constitutes a "failure" or an undesirable event that warrants the recording of a line in the HRR. Without a clearly defined success criterion, any attempt to evaluate risks that matter in a considered system will be misleading at best. For example, in a GHZ RA, the success criteria for a rockfall study were defined as: no casualties due to rockfall. However, as zero risk does not exist, the success criteria should be defined probabilistically: success is reducing a casualty probability to 10^{-5}/year or lower.

The success criteria should always be complemented by a measure of the tolerance. If the tolerance covers a range of possible consequences (e.g., from 1 casualty to 100 casualties) coupled with their tolerated likelihood, then a tolerance criterion is defined, a notion we will develop in Section 4.5. Success criteria may become rather complex, depending on the aim of the assessment. A more sophisticated, four-dimensional success criterion could be suggested as follows: (a) reduce casualties as described above, (b) reduce closure time of the path at the toe to max 1 day/month, (c) adapt to climate changes (runoff surges on the slope), and (d) survive for at least 5 years without the need for capital expenditures and extraordinary maintenance.

http://dx.doi.org/10.5772/66066

2.3. What is a risk assessment: do pigs fly?

The ubiquitous common-practice RA methods are FMEAs and Probability Impact Graphs (PIGs, see **Figure 1**). Due to common mistakes described above (and further down in this chapter) and their misleading nature, these methods do not allow to grasp the full true story of the multihazard (or convergent) RAs that should be an integral part of responsible DMs lives. PIGs present at best a colorful chart, usable as a rough first estimate [6], but not sufficient for complex decision-making for critical infrastructures of our modern society. People often accept PIGs uncritically and trustingly until something goes wrong. As shown above, problems arise from the use of an unclear glossary, the basic structure of the HRR, and continue with simplistic definition of probabilities and censored consequences. Experience has shown that PIGs often end up with a major confidence crisis, possibly leading to societal or regulatory opposition [7–9].

2.3.1. Common practice

Misstep example: PIGs are usually drawn with symmetrical color schemes (**Figure 1** is symmetrical around the yellow colored diagonal), or almost symmetrically, in such a way that high consequence, low probability events have the same risk "color" as low consequence, high probability events (the extremes of the yellow diagonal in the case depicted in **Figure 1**). Following that scheme leads, for example, to prioritize the risk of an asteroid obliterating your house (and family) in the same class as you getting a cold, a prioritization that would be considered misleading by most readers and DMs. Symmetry implies that a "Fukushima scenario" (catastrophic impact, very low probability) is considered equal to your next flu (low consequence, high probability).

Erroneous statement: The matrix cannot be colored in ways that lead to misleading prioritization. Symmetrical coloring is a clear flag for erroneous statements in the interpretation based on the prioritization.

Rule 5: When using a risk matrix (Probability Impact Graph, PIG) for the risk prioritization (usually stated as a specific color), there is the need to check that the colors "match" real-life (corporate or societal) expectations.

Quick fix: Use extreme cases to see if the coloring scheme still makes sense. If you really have to use the PIGs representation, alter the coloring scheme until it makes sense. Do not try to guess a tolerance threshold: there are specific studies required to develop a defensible one.

Correct: Matrices cannot be symmetrical. Consider all extreme cases and if you really feel like "coloring the cells," pay attention to what the colors will tell your users.

Misstep example: Develop classes for the consequences and probabilities for your risk prioritization. It is not by adding another color or class that you will solve the binning systemic error.

Erroneous statement: In order to increase "precision" of the studies, the Geologic Bureau asks all permits to be developed using a 5 × 5 matrix (1 = negligible, 2 = small, 3 = medium, 4 = large,

5 = catastrophic) instead of the 3 × 3 matrix (1 = small, 2 = medium, 3 = large) used to date. It would be interesting to ask them how they are going to compare "old" studies with "new" ones, and what 1 through 5 mean today when compared to 1 through 3 yesterday.

Rule 6: Do not bin.

Quick fix: Above all, avoid using indices. Stay quantitative. The math is simple, and the RA will be tremendously improved if your "risk-dots" are in the proper p, C position and not just binned in. This fix will allow for rational prioritization but also enable providing insurance-limits computations and avoid being overwhelmed by bin-overcrowding.

Correct: You do not need to develop classes and indices; use "real" ranges of values: probabilities vary between 0 = cannot happen and 1 = certain to happen, and each one of your *hazards* can be allotted a range (See Section 3.1). *Consequences* range between 0 and infinite (in a metric we will define in Section 3.2) and you can evaluate a range for each hazard record. Do not color "cells," as you will soon be able to develop a tolerance level in Section 4.5.

2.3.2. What is needed

A recent decision by an Environmental Review Board in Arctic Canada [10] quoted the following five requirements for a socially and technically acceptable (GHZ) RA.

1. Compilation of a proper glossary containing a description of all the terms used in the project and its development, especially those that might have a common use, which differs from the technical meaning (such as "risk," "crisis," "hazard") in compliance with ISO 31000.

2. Definition of the project context in compliance with ISO 31000, including all the assumptions on the project environment, chronology, etc.

3. Properly defined HRR covering:

   - Clearly defined system of macro- and subsystems/elements and their links describing for each one of them: (a) expected performances, (b) possible failure modes, (c) quantification of the related ranges (to include uncertainties) of probabilities evaluated as numbers in the range 0–1 (mathematical characterization) with a clear explanation of the assumptions underlying their determination, and (d) associated magnitude of the hazards and related scenarios.

   - An independent analysis of failure/success objectives.

   - A holistic consequence function integrating all health and safety, environmental, economic and financial direct and indirect effects.

   - Applicable published correlations and information.

4. RA is expected to use a unified metric showing consequence as a function of all health and safety, environmental, economic, and financial direct and indirect effects. This will be done in a manner that allows transparent comparison of holistic risks with the selected tolerability threshold (see below).

http://dx.doi.org/10.5772/66066

5. Consequences will be expressed as ranges, to include uncertainties. When evaluating the consequences, the RA will:

   - Explicitly define risk acceptability/tolerability thresholds, in compliance with ISO 31000 international code. These will be determined in consultation with potentially affected communities, using a unified metric compatible with the one described above for consequences.
   - Risks and tolerability or acceptability will be developed separately, in such a way not to influence or bias judgment of the assessors or evaluators. Risks will then be grouped into "tolerable" and "intolerable" classes. The risks in the intolerable group will be ranked as a function of their intolerable part. Mitigation efforts will be allotted proportionally to that ranking.

We will now proceed with a systematic discussion of the requirements and their meaning.

3. Characterizing risks

In the ANPcene, insurers are facing "new" challenges when insuring against GHZs, especially those caused by human activity. They have realized that because of the dynamic evolution, the usual actuarial point of view on risk faces significant challenges and can be misleading. The indiscriminate use of force majeure (FM) and insurance denial to protect themselves is actually detrimental to their business and their clients. What an epiphany! Looking only in the rear-view mirror while driving is indeed going to complicate the steering of the vehicle! Now, insurers have always worked like that, i.e., using past data (statistics) to evaluate their risks and business opportunities, and they have already got their share of misery from climate changes and other GHZs. GHZs caused by ANPcenic fast-track evolution are typically an arena where using actuarial data and statistics can only be wrong and expose everyone, including the insurers, to enormous risks overexposures.

Unfortunately, insurers have asked hazard specialist (geoscientists, weather people) help in solving their conundrum, a mistake we oftentimes see occurring in various business spaces. As mentioned earlier, hazard specialists want to measure what they know, but they often confuse hazard with risks, and by managing hazards instead of risks, they end up being ineffective or inefficient, i.e., squandering money, not getting results, or leading to unjustified insurance denial. Rule 1 is of paramount importance in mitigating this.

3.1. Evaluating probabilities, frequencies, their relationship

In this chapter, we do not discuss qualitative evaluations as we consider them not aligned with modern societal needs of transparency and rationality.

We start this section by looking at some particular values, which are often encountered in the hazard-risk literature. The first one is the "Act of God (AoG)," a term that entered in contract practice more than a century ago to indicate unforeseeable and unmanageable events such as hurricanes and earthquakes, and thus trigger the activation of FM clause. The definition of an AoG can be understood with an example: if we consider a commercial contract for a facility in Salt Lake City, one could have said that a tornado in Salt Lake City was an AoG (scientific consensus that a tornado was not a credible, actually "impossible" event in Salt Lake City) until one happened on August 11, 1999. From that point on, a tornado in Salt Lake City became a rare, but credible, event and hence, should not be considered an AoG anymore.

Of course at this point, terms such as "credible event (CE)" have to be clarified, as their definition is intimately linked with the limiting value of FM.

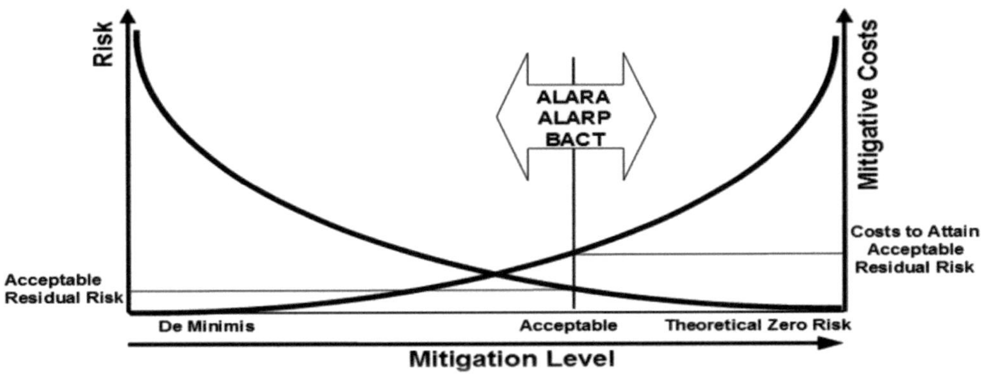

Figure 2. Risk reduction vs. mitigative costs (investments), "acceptable" mitigative thresholds, and acceptable residual risks. Zero risk is a theoretical concept only.

Continuing our discussion, in recent times, "standardized levels of risk reduction/mitigation" have been formulated (in various industrial spaces), and at least three of these definitions are now in common use among analysts (http://www.riskope.com/2016/07/07/geohazards-probabilities-frequencies-and-insurance-denial/). These three levels of risk mitigation also represent a convenient way to elude explicit tackling of risk tolerance (see Section 4.5), especially when the delicate theme of human life has to be dealt with. However, these standardized levels of risk reduction can also be seen as a way to define "the state of the art" practice. Anything below a predefined ALARA, ALARP, BACT level can be considered as "negligent" (NB: in recent years, it has been noted that public opinion tends to consider negligent mitigative levels that are above these limits, thus leading to image risks even when "reasonable mitigative behavior" is followed by corporations, governments) (**Figure 2**).

Here are the definitions of the three "standard" levels of mitigation:

ALARA: ALARA means as low as reasonably achievable [11] and can be used for rockfalls and landslides along roads.

ALARP: ALARP stands for as low as reasonably practicable, a term often used in safety-critical and high-integrity systems. For a risk to be ALARP, it must be possible to demonstrate that the cost involved in reducing the risk would further be grossly disproportionate to the benefit gained. It is a best common practice of judgment of the balance of risk and societal benefit. It could be used for GHZs impinging on critical infrastructures.

BACT: BACT stands for best available control technology. It can be used for toxic dumps and tailings storage facilities.

No matter which mitigative level is adopted as "standard, nonnegligent, practice," a fundamental step is to define against which event it is necessary to mitigate (as it is a "credible" event) and against which event we humans have to humble, as it can be considered an AoG.

Like for risk, the technical definition of credibility differs quite substantially from the "colloquial" definition. Industries where major accidents/events are a concern, generally define credible accidents as accidents within the realm of possibility (i.e., probability higher than 10^{-6}/year) and with a propensity to cause significant damage (at least one fatality). This concept comprises both probable damage caused by an accident and probability of its occurrence. As the threshold value of 10^{-5} is a generally accepted value used in seismic, geological, and other sciences to define "maximum CEs (MCE)," it can be assumed, for the optimization of the FM clause, that events with a probability of occurrence of less than 1 in a million (10^{-6}) or less belong to AoG, whereas events with a probability of at least one in hundred thousand (10^{-5}) or more are credible and more importantly, although rare, foreseeable, and should be mitigated.

As mentioned earlier (Section 2.2.2), just around the time of the 2008 recession, the "Black Swan (BS)" buzz-word phenomenon exploded. It caught the imagination of many, but in our mind, it brought forward a slanted image. Let's explain the point: BSs were originally defined [12] and currently defined on the web as "an unpredictable or unforeseen event, typically ones with extreme consequences." Unpredictable can be assumed to be beyond credibility or at AoG level, hence one in a million (10^{-6}), or less. Interested readers can study the Blackett report [13] to find a balanced set of approaches to low probabilities risks. Thus, when one reads that "geopolitical BS events, such as the Arab Spring and the Japanese earthquake, have further complicated the market dynamics," we are reading about some illogical statement. The same when we read about tailings dam breaches being considered BSs: they cannot be as their rate of failure is 10^{-3} to 10^{-4}, at least two orders of magnitude more likely than an AoG and definitely in the credible range. In addition, if major economic turmoil occurred 17 times in the last two centuries (rate of occurrence is $17/200 = 0.085$/year), it is also difficult to consider it a BS. However, societies have very short, selective, memory, so we all thought that 2008 was one unpredictable and unforeseen event despite such a high recurrence of similar events. That's WRONG! Furthermore, social scientists [14] have argued that major disasters do not occur "out of the blue" but incubate over a period of time with potentially identifiable patterns of attributes. The thesis was not new, and its germinal concepts started between the thirties and the seventies, with systematic studies of industrial accidents conducted by various insurance companies. We have found in our day-to-day practice numerous cases where GHZs seem develop following this "incubation" principle, provided a serious record of "near misses" is maintained.

In the case of the Fukushima earthquake, reportedly the sea defences were designed for the MCE, but there are also voices that *a priori* the height of the tsunami at Fukushima might have been regarded as inconceivable. However, if our lack of knowledge about such rare events had been admitted, then, in view of the spectrum of potential consequences (uncensored and unbiased spectrum, of course), there should have been an incentive to ensure that the design was robust to the lack of knowledge. Reportedly, designers trusted only one line of defence (hence, the project was not robust or reliable, but fragile), leaving the electrical commands ready to flood in the underground of the plant (yet, as engineers, we know that trusting one line of defence, the properties of one material, device, etc. is not good sense). Considering Fukushima a BS was/is dead WRONG, and the resulting consequences due to the dismissal of the uncertainties became apparent.

"Availability heuristic" [15] is a very well-known human cognitive bias tainting decisions under uncertainty. That bias can explain why the 2008 recession was considered unheard of, a BS: just because most people did not remember (were not even born) in 1929! The BS "fad" is indeed based on humans having "short memory" and considering the last events as "unique." Sometimes, we are forced to use availability heuristics because available data are indeed very scarce and only recently gathered, but reliable statistical evidence will systematically outperform "intuition" when "looking backwards" in time to past events to draw conclusions. Looking backwards, however, is not enough and actually it is critically limiting and incomplete when we are confronted with managing risks. A good RA, especially a GHZ one, has to be "looking forward," examining "classic" scenarios and hypothetical ones that have not yet occurred (or not yet occurred with larger magnitudes) before making management decisions. As a matter of fact, Kahneman and Tversky [15] have explored in detail how human judgment can be distorted when making decisions under uncertainty: humans tend to be risk-averse when facing the prospect of a gain, and paradoxically risk-prone when facing the prospect of a loss (even if the loss is almost certain to occur)! Thus, using improper methods like PIGs (See **Figure 1**), which will almost surely lead to confusion, losses, and poor planning, sits well with "main-stream" human nature.

Thus, we can state that the range of the probabilities we need to define spans from 10^{-6} to 1, so from the threshold of credibility to "certainty." In performing probability estimates, we need to remember that it is better to be roughly right than precisely wrong, meaning that uncertainties should always be part of the estimates and be explicitly stated. Probabilities can be evaluated through probabilistic models, statistics, and direct encoding of expert knowledge. The detailed discussion of these methodologies is not within the scope of this chapter but can be found in technical literature [16, 17].

Misstep example: Giving one precise value for the probability.

The past can never be assumed to equal the future. At best, it can be used as a point estimate.

Erroneous statement: Based on the historic series of flooding in this area, it is assumed that flooding to +4 m will occur every 50 years.

Rule 7: Always consider a range of probabilities in order to include the range of uncertainties.

Quick fix: Uncertainties will always exist. Consider the limits of our human capability to estimate events. Give one pessimistic probability, usually Common-Cause Failure based (i.e.,

http://dx.doi.org/10.5772/66066

in the case all redundancies fail because of a common flaw), and one optimistic probability with the foreseen mitigation active. If probabilities are transparently considered uncertain, then a Bayesian update mechanism can be implemented when new data become available.

Correct: Based on the flooding historic series in the area, it is assumed that flooding to +4 m has a yearly probability ranging between x and y. Thus, it is possible to evaluate (Poisson process) that the probability of occurrence of 1, 2, 3 floods at +4 m is x1–y1, x2–y2, and x3–y3.

3.1.1. Quantitative by analogy (applied to L'Aquila earthquake)

NB: Data for this summary have been gathered through media and publicly available records. Details we consider "irrelevant" to this discussion have been omitted because of space limitations.

In 2009, the city of L'Aquila, located in Italy, was hit by an earthquake. The city featured many historic public buildings and antique residential structures, which had not been retrofitted, Italy being a country where retrofitting of old (privately owned) structures to meet new seismic safety criteria is reportedly not enforced (decree OPCM 3274/03, art. 2, comma 6). A 1999 study on the vulnerability of public, strategic, and "special" buildings had indeed shown critical vulnerabilities. The area is seismic as witnessed by major earthquakes recorded since the year 1349, then 1452, 1461, 1501, 1646, 1703, 1706, 1958, and 2009 (eight major events in approx. 650 years, or 0.0123 (1.23%) per annum, excluding the 2009 event). This last quake led to 309 casualties, 1600 wounded (200 very severely), 65,000 evacuated out of the city, and damages for over 10B€. Prior to the tragic event, a "swarm" of foreshock earthquakes with almost 100 times the average rate was recorded in and around L'Aquila. The swarm triggered a crisis status due to public's panic, further fueled by independent scientists' opinions. The swarm had started in December 2008 (magnitude, M = 1.8), then in January with M = 3 gradually and continuously evolving with increasing intensity and frequency to the date of the major event.

For the sake of our discussion, that 1.23% per annum, accelerating very significantly and increasing in intensity has to be placed in comparison with other phenomena: the world portfolio of hydraulic dams has a rate of failure near credibility (10^{-5} to 10^{-6}), tailings dams major breaches at 10^{-3} to 10^{-4}. There was clearly something significant occurring and an official government body called the National Commission for the Forecast and Prevention of Major Risks had six top officers participating in a meeting with the public on March 31, 2009, 6 days before the nefarious earthquake (Magnitude 6.3/Richter 5.9), and a day after the latest, and strongest event in the swarm.

The team spoke directly with the public rather than via the civil protection department. The public's concerns were entirely dismissed, and as a result, some of the town's residents changed their behavior of seeking shelter outside, as they were used to do when tremors happened, staying indoors instead.

The team was brought to trial for manslaughter in September 2011 for the advice they gave in that meeting.

Scientific American [18] rightly wrote that "this was not a case against science, the judge recognized the unpredictability of such an event already in the indictment, but a judgment against the failure of scientific communication (of risks)."

It is time that geoscientists, seismologists, and engineers, who are very capable and respectable hazard specialists, recognize that RAs are an area requiring specific knowledge. RAs should be prepared by risk specialists and hazard knowledge constitutes at most half of the equation.

Operating by comparing the long-term rate of occurrence, accelerated frequencies, and using analogies with other catastrophic occurrences (dam breaches), would have helped to rationally characterize the ongoing changes and the uncertainties. Armed with that knowledge, it is likely that the communication to the public would have been different. Reportedly, all the second instance tribunal ended up releasing the accusations. It is not our role to discuss the judges sentencing.

3.1.2. Quantitative by model(s)

A large alpine landslide was the object of a quantitative RA, which used a probabilistic slope-stability method [19] to show which failure modes were most critical in terms of probability of sudden accelerations (paroxysms) and consequences to the transportation corridor and the watercourse lying at the toe (**Figure 3**). Failure modes varied from relatively shallow slides (<500,000 m^3 in volume and less than 10 m deep) possibly occurring on top of the massive historic slow-creeping mass (10 Mm^3, up to 100 m deep).

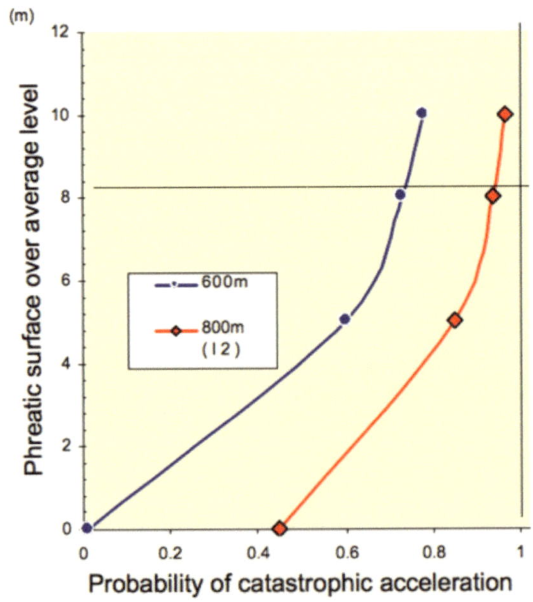

Figure 3. Probability of catastrophic acceleration vs. phreatic surface over average long-term level at two locations determined using a probabilistic slope-stability method [19].

Based on the evolution of the topographic displacements, a model was built and maintained to evaluate the potential runoff of the masses; hence, the consequences of failures on the surrounding infrastructure [20]. Monitoring revealed that the slide responded as foreseen by the probabilistic analyses in the case of a particular cycle of adverse meteorological events. All stakeholders were shown the evidence of the instrumental and probabilistic analyses together with the expected landslide evolution. Thanks to clear risk communication and understanding, consensus was reached on the appropriate level of mitigation in the form of a drainage tunnel, which gained strong "social" support and was built.

3.1.3. *Quantitative by internet of things*

Data management systems linking multitemporal objective data acquisition with dynamic convergent RA platform are needed to:

- capture data from many sources (manual, automatic)
- create a common database for all—no disagreements about whose data to use
- present the data in an intelligible form via the web-based portal
- prepare reports, with up-to-date data and analysis, virtually automatically
- generate efficient meetings and strongly reduce "emergency" clarification meetings
- spare time spent chasing for factual information—decision can be made more rapidly with agreed data
- allow management at all levels to hold paperless meetings with current data viewed and analyzed on the system
- give access to project's historical data to all stakeholders from one unified data library
- make data available across all phases of endeavors and across all stakeholders and all (authorized) parties
- manage alert/alarm/action levels sending email and/or SMS to controlled users groups.

The link between the data acquisition (sensors, monitoring stations, etc.) and the RA platform should use Bayesian updates of probabilities, frequencies, and other selected parameters to distill the data used in the RA. Connecting a dynamic quantitative risk-analysis platform with a high-performance data-gathering technique reduces costs, avoids blunders, and constitutes a healthy management practice, especially for long-term projects requiring short- or long-term monitoring (**Figure 4**).

This allows all parties using the system, including senior managers and the next generation of managers to easily and rapidly review the current status of the project and start management procedures when preselected thresholds are reached. It allows management and engineers to focus on RM. There are tangible and intangible benefits deriving from big data use. Intangible benefits are considered to be potentially as valuable as the tangible (directly quantifiable) benefits and include a focus on hazard identification and dynamic risk evaluation rather than (uninformed) risk taking, as well as the use of one dataset visible to all parties.

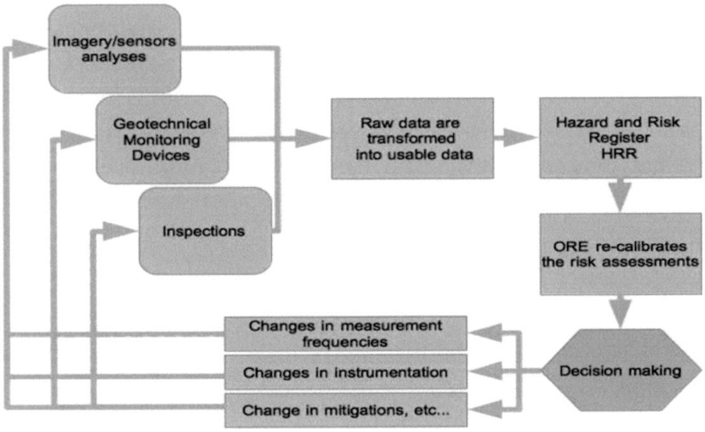

Figure 4. Convergent, scalable, dynamic scheme for a RA/RM platform like ORE (Optimum Risk Estimates, ©Riskope).

3.2. Multidimensional consequences analysis in the Anthropocene

In a risk management model, it is essential to have a vision of the losses that an event could cause. In too many cases, risk studies take into account losses in a limited way, either by ill-will or by misunderstanding of the real implications of an accident or event.

The need for a unified, emergency/accident scale is vital to facilitate clear communication and mutual understanding of the nature of the emergency, by the public, government agencies, and responding organizations. It has been stated that "50% of the problems with (risk) communication are due to individuals using the same words with different meanings. The remaining 50% are due to individuals using different words with the same meanings" [21]. In many countries, legislation still has not provided definitions of "disaster" or "emergency," as well as the difference in impact and immediacy of response. An objectively calculable emergency scale should, therefore, quantify and clearly communicate the notion of "emergency."

The elements that must be considered for the definition of the overall metric of losses are, at least and not in any particular order: (a) direct & indirect, (b) health and safety, (c) environmental, (d) image and reputation, (e) legal, etc.

In some national scale RAs, we have reviewed that the aggregated extent of damage is calculated by converting each damage into the same unit, i.e., monetary value.

The marginal costs are equivalent to the approximate amount of money that the society is willing to pay in order to reduce the extent of damage of an indicator by one unit [22]. That approach is simple and allows for specialist discussions. The public requires a more direct approach. **Figure 5** displays the results of a GHZ RA which uses a sophisticated multidimensional monetary scale to express a consequence metric as well as words.

Solutions have been proposed [23, 24], recognizing the public needs to be well informed with accurate, time critical information, especially in the aftermath of a catastrophic event, including, of course, GHZs. Primary information source are generally event-specific scales that are inconsistent in their categorization and measurement, adding confusion to public responsiveness. Furthermore, these scales are not extendable to new emergencies in a changing world. Society reportedly needs the development of a unified emergency scale to facilitate communication and understanding. Such a scale could inform local communities with regional community-specific information and could be extendable for further use by professional responders. Research in the reference above [23, 24] elicited 15 dimensions of an emergency using a Delphi-like process and then ranked the dimensions by importance utilizing Thurstone's law of comparative judgment.

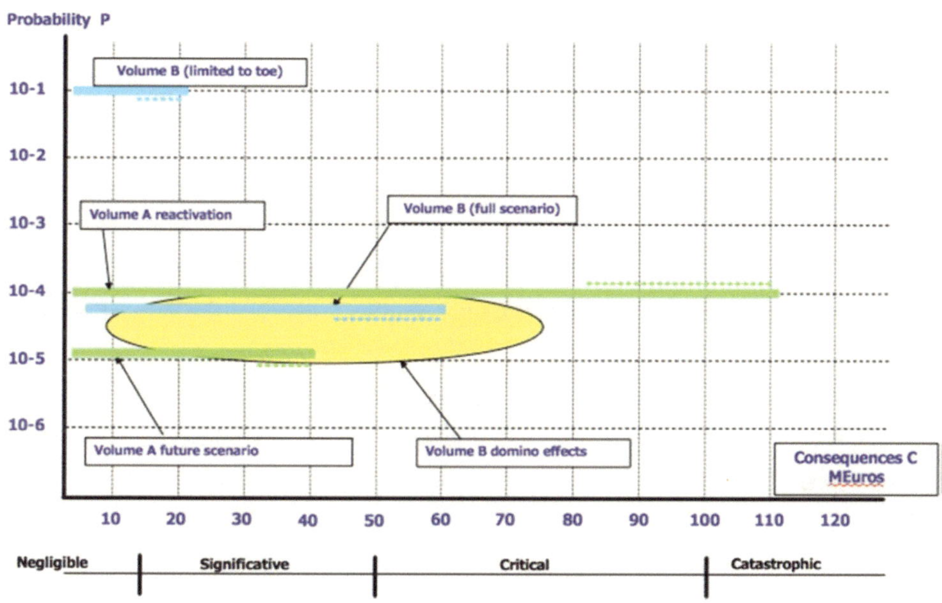

Figure 5. A quantitative result of a GHZ (various scenarios) RA where consequences are expressed with a sophisticated multidimensional monetary scale as well as words.

4. Real-life Anthropocene accident consequences and risk perception

When making choices, individuals, societies, and governments are driven by their needs and their perceived or factual uncertainties and hazards. Thus, different socioeconomic, cultural, and religious contexts will need different selections even within a same jurisdiction, over time. The tobacco-industry debate, the climate-change debate, and the nuclear- and mining-industry one are all examples of that multifaceted reality. People in need will often forget or disregard hazards. In countries like Cambodia, plagued by land mines and unexploded ordnance, demographic pressure (and the specter of starvation) drives, for example, people to cultivate perimeters considered as hazardous. Governments will often ponder costs and benefits of large projects, often based on incomplete and misleading evaluations.

Misstep example: Giving one precise value for the consequences. The human brain is generally good at imagining the best and the worst scenarios, but we see many times that people censor the range considered. In modern society, he who hides risks dies, sooner or later.

Erroneous statement: If we build a plant in this location, the consequence of an explosion could be that 50 nearby residents die.

Rule 8: Always consider a range of consequences.

Quick fix: Uncertainties will always exist. Don't censor!

Correct: If we build a plant in this location, the consequence of an explosion could be that 10–100 nearby residents die, their residences are destroyed between 20–50%, there will be jobless people for 1 to 2 years, etc., and the site may be contaminated by chemicals in a radius of 2 km, causing 100 M$ to 200 M$ clean-up costs.

Misstep example: The consequences of a small car accident are that you arrived late AND you have some repair to make AND you might be bruised. Why is it that consequences of a facility evaluation often consider only the "worst" among, for example, H&S, environmental, or BI?

Erroneous statement: Often seen in FMEA/PIGs applications: use the largest consequence among H&S, physical losses, and environmental to "bin" the risks.

Rule 9: The consequences are almost always a mix of those associated with health and safety (H&S), business interruption (BI), environmental, etc., at least in an additive way.

Quick fix: Record all types of consequences and then work with a blended metric.

Correct: Define, a priori, a multidimensional cost of consequence function where the various components are added to obtain the total.

4.1. The German model

German researchers in the field of theoretical risk management developed a series of metaphors to describe public perception of risks [25]. They are summarized below, with examples.

Sword of Damocles has very high-potential consequences, paired with very low probability. Nuclear power plants (Fukushima), large-scale chemical facilities (Bophal, Seveso), hydro-dams failures, and meteorite impacts are typical examples.

In **Cyclops,** it is only possible to ascertain either the probability of occurrence or the extent of damage, while the other side remains uncertain. A number of natural events such as volcanic eruptions (Vesuvius), earthquakes (various "large ones" like San Francisco, Tokyo, etc.), and floods belong in this category.

Pythias includes risks associated with the possibility of sudden nonlinear climatic changes such as the risk of self-reinforcing global warming or of the instability of the West Antarctic

http://dx.doi.org/10.5772/66066

ice sheet. The extent of damage is unknown, and the probability of occurrence cannot be ascertained with any accuracy.

Pandora's Box has strong uncertainties in the probability of occurrence and extent of damage (only presumptions) coupled with high persistency. Beside persistent organic pollutants, biosystem-changing endocrine disruptors can be quoted as examples.

Cassandra is characterized by a relatively lengthy delay between the triggering event (e.g., nuclear radiation exposure below a certain critical threshold) and the occurrence of damage. This case is naturally only of interest if both the probability and magnitude of damage are relatively high. In other cases, it gives a false sense of safety.

Medusa refers to the potential for public mobilization. This criterion expresses the extent of individual aversion to risk and the political protest potential fueled by this aversion, both of which are triggered among the lay public when certain risks are taken. This risk class is only of interest if there is a particularly large gap between lay risk perceptions and expert risk-analysis findings. Some innovations are rejected although they are hardly assessed scientifically as threat (cell phones, high voltage lines, etc.).

In **Table 1**, we have attempted to relate "knowledge level," "degree of incertitude," "main criteria," and finally the German metaphor name.

Knowledge level	Degree of uncertainty	Main criteria	Metaphors
Minimal	Ignorance	Probability of occurrence and extent of damage are highly unknown to science	Pandora's Box
Fair	Uncertainty	Probability of occurrence or extent of damage or both are uncertain (because of natural variations or genuine stochastic relationships)	Cyclops Pythias Cassandra
High	Known distribution of probabilities and corresponding damages	Probability of occurrence and extent of damage are known	Sword of Damocles Medusa

Table 1. A summary of the German metaphors.

We oftentimes see different accident types, with identical or very similar single-accident direct consequences but very different global impacts, generate surprisingly different public reactions. Would the metaphors help explain why?

4.2. Identical consequences generating different behavior

Indeed, identical single accidents' consequences can lead to diverging societal impacts. **Table 2** gives a list of well-reported historic flooding examples around the world with identical or very similar single-accident direct consequences. The global impacts were very different and generated surprisingly different public reactions, in particular with the Stava tailings dam breach.

Casualties	Location	Year
230	Marrakesh flash flood, Morocco	1995
235–244	Philippine Floods, Philippines	2009
246	Rio de Janeiro floods and mudslides, Brazil	2010
268	Val di Stava dam disaster, Italy	1985
299	Nagasaki, massive rain and landslide, Japan	1982
300	Quebrada Blanca canyon, landslide, Colombia	1974
313	Jambi, Batanghari, Tondano, Indonesia	2003

Table 2. Historic flooding examples.

Let's examine some examples:

Accident a1) Floods kill hundreds of people around the world (NB: Floods in Virginia, Texas, and China in 2016 have also killed dozens/hundreds). There is short-term commotion but of limited impact. Societally, a1 has such a high rate that it is considered a "fact of life" by society, whose perception becomes numb, while the public is not informed and goes by totally unaware. However, a1 could mobilize public opinion, because people could feel "surprised" and almost "betrayed" by scientists and public officers. At that stage, a1 could be interpreted as a Medusa "freeze," a public stupor before outcry, and mitigative actions would be decided on the spot (building new dikes, etc.). As risk perception depends on the viewers' position, for the population potentially exposed (near to riverbeds), this risk is a Sword of Damocles. However, as human often believe they are different from others, accidents only occur to others, so personal likelihood is perceived as very low.

Accident a2) A single accident of a Tailings Dam with "identical" consequences in terms of Casualties, but probability (rate of occurrence), most likely very different (a1 >> a2 as p(a1) is more than 1% per year, p(a2) is around 1/1000 or lower per year). Societally, a2 is a Cyclops: it is easy to imagine that the exposed population will die, but the probability of the accident is highly uncertain because of extant, apparently sufficient, safety rules (See Samarco Dam breach in Brazil, 2015). Death is an "expected" consequence, but people think its occurrence is most uncertain. If an accident occurs, there is an immediate awareness, in some cases panic, due to the fact that at least that occurrence is now certain to have happened. The case gets lots of attention and mitigation/punishment of the guilty is decided, but there is likely no Medusa "freeze." For the potential victims, this risk is a Sword of Damocles, like for a1.

4.3. Accident perception and crises potential

We are now ready to discuss accident consequences and risk perception using various ANPcenic nefarious events (not necessarily all belonging to the GHZ family, but useful, nevertheless, for the sake of discussion).

http://dx.doi.org/10.5772/66066

Accident c1: Class 5+ nuclear accidents (Fukushima, this was a GHZ linked accident, Tschernobyl, Kyshtym disaster, Windscale fire, Three-Mile Island accident, First Chalk River accident, and Lucens partial core meltdown) were likely to be interpreted as Sword of Damocles metaphor but would likely be Medusa now.

Accident c2: Nuclear-waste storage accidents (possibly due to GHZs in the future) can be easily considered to belong to the Pandora interpretation, evolving into Cassandra (presently), and then later they may evolve into Cyclops.

Accident c3: Post-accidental exposure is Cassandra, but for people onsite, it is a Sword of Damocles.

We can now set up **Table 3** with various accidents' examples and their related metaphoric descriptors at personal/local or societal/general level as follows, including public perception and likely reactions. It appears from **Table 3** that when personal/local risks belonging to the Cyclops, Sword of Damocles metaphors has societal/general potential to be finally perceived as Medusa, and then they are to be considered as "societally intolerable" because they soon trigger mitigation, moratoria, protests, etc. In conclusion, the German metaphors can be used for discussing image and societal perception of risk scenarios.

Accidents and related GHZs risks	Personal/local metaphor	Societal/general metaphor	Public perception
a1 natural flooding	Sword of Damocles	Medusa	Intolerable, leads to crisis if awareness raises, sense of betrayal
a2 Dam breaches/man-made accidents	Sword of Damocles	Cyclop	Mitigative measures can be decided
c1 Class 5+ nuclear (including GHZs generated ones)	Sword of Damocles	Medusa	Intolerable, leads to crisis if awareness raises, sense of betrayal
c2 Nuclear underground storage (including GHZs generated ones)	Pandora to Cassandra	Cyclops evolving to Medusa	Mitigative measures will be proposed, not necessarily implemented until the Medusa stage is reached, then there will be a crisis.
c3 Post accident exposure of nuclear underground storage.	Sword of Damocles	Cassandra evolving to Medusa	Mitigative measures will be proposed, not necessarily implemented until the Medusa stage is reached, then there will be a crisis.

Table 3. Various accidents examples and their related metaphoric descriptors.

If any scenario is considered to have a high Medusa perception, it can also be considered to be at the limit of societal tolerance, whatever the factual consequences may be, even if they are relatively small. If the Medusa is considered of fast development, then the ensuing crisis will probably lead to a major crisis, with potential catastrophic corporate/governmental consequences.

4.4. Third-party Hazards' possible impacts

As mentioned later on (See Section 4.5.2), third-party operational hazards can generate risks requiring strategic shifts. We live in a complex interconnected world sometimes generating very difficult to detect/understand interdependencies. These are of different kinds: *physical, geographical, logical, and cyber interdependencies.* Of course, the last two are not directly linked to GHZs.

Example of physical and geographical interdependency: Following recent reports from various sources, it seems that fracking extraction techniques have the potential to generate seismicity; climate change is reportedly increasing the rate of extreme events in various areas of the planet.

Example of geographical interdependency: An accidental spill from a "third-party" dump in an area may result in a severe restriction of activities (exclusion zone) for other activities.

Based on the above examples, it is easy to see that an HRR can stop at the "property perimeter," which does not necessarily correspond to the "risk battery limit," but doing so corresponds to an "ostrich" stance, that is, to purposely censoring the scope of the study, leaving open vulnerabilities to risks that may affect strategic goals.

Neighbors' (third parties') operational hazards can generate risks impinging on operations: it is enough to "be there," and the image and reputation may even suffer a blow from those hits. Some of the generated risks may be intolerable, or even unmanageable, thus requiring a strategic shift (See Section 4.5.2 below). Only a rational and well-balanced RA can help decide and avoid the pitfalls.

4.5. Risks and risk tolerance

Tolerable risk curves (tolerance thresholds) are always project- and owner-specific and indicate the level of risk, which has been deemed acceptable by an owner for a specific project or operation (possibly taking into account public opinion). This means, as an example, that within large companies, corporate-risk tolerability may differ quite substantially from a branch operation's tolerability. The development of empirical-estimated tolerability curves requires caution and continuous calibration; they should always be defined by a group and not by an individual.

In an RA, great attention must be exerted in ensuring that the acceptability curves are derived for the considered risks: curves derived for hazardous industrial activities cannot be used for GHZ risks or business risks.

The first explicit examples of Risk Tolerance/Acceptability criteria were published in the mid-eighties [26]. For example, in more recent times, the Australian National Committee on Large Dams [27] also published its own criteria (**Figure 6**). In general, two criteria are defined, a prudent or risk-averse one, thus a low-bound curve, and a risk-prone aggressive, high-bound curve.

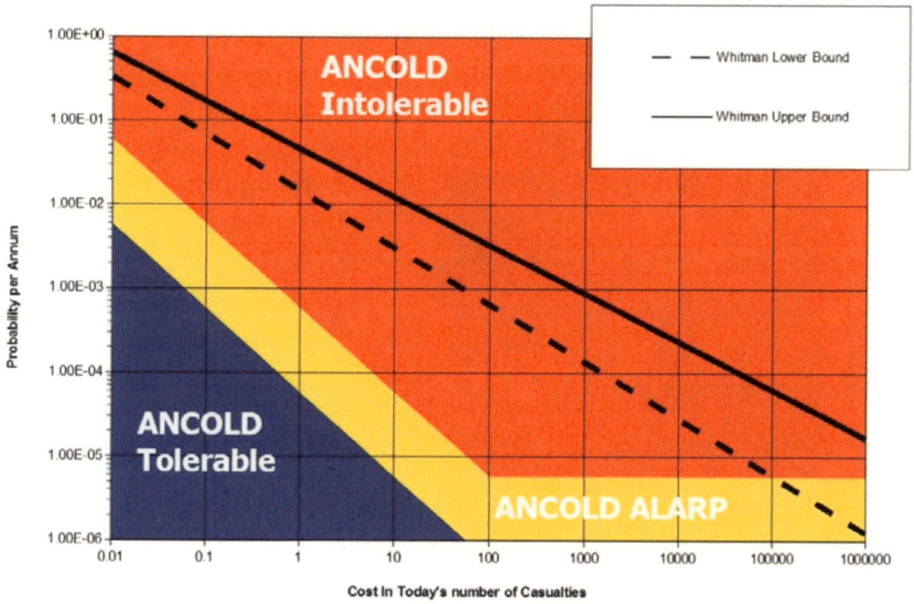

Figure 6. ANCOLD and Whitman published tolerance thresholds [26, 27].

4.5.1. Defining tolerance

"Instinct" and "intuition" are often poor advisors when attempting to define risk tolerance thresholds to replace the arbitrary PIGs (risk matrix) coloring. Such intuition-based thresholds constitute an attempt to diverge from the classic "diagonal-color scheme," but the minimum consequence and maximum probability oftentimes show again the same risk priority as minimum probability and maximum consequence (Remember Rule 5, Section 2.3.1).

Misstep example: Coloring schemes or thresholds' criteria mismatch with accepted thresholds. We have seen RA rejected because they were not defensible at that level (**Figures 6** and **7**).

Tolerability has to be defined in order to allow proper decision-making.

Tolerability definition requires transparent communication with stakeholders.

Erroneous statement: Tolerance is a single line that, if crossed, is deemed intolerable.

Rule 10: Use published societal tolerance thresholds (**Figure 6**) to see where you are standing and develop your own tolerance criteria for corporate affairs.

Quick fix: Do not use prefabricated PIGs with arbitrary cell-limit definitions or arbitrary colors. You do not need the cells! And you can add to your plot a well thought-out tolerance limit.

Correct: Link every probability to the consequences at stake to define an empirical tolerance threshold and then compare it to publicly available literature such as **Figure 6**.

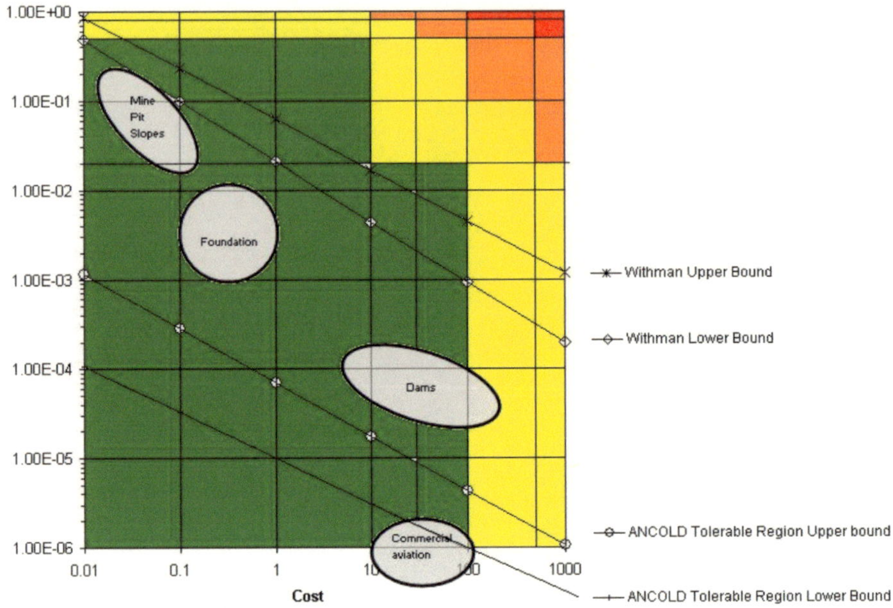

Figure 7. A colored risk matrix superimposed over ANCOLD and Whitman's thresholds. Vertical axis: probability, horizontal axis: cost (consequences).

Figure 7 shows a real-life, colored risk matrix from a corporate RA superimposed over ANCOLD and Whitman's thresholds. One can see that the green-colored cells are in the intolerable part of both Whitman's and ANCOLD, which, in the case of a mishaps, would prove difficult to defend corporately, societally, and legally. One day, such a case will be challenged in the court of law: will it be negligence, or misrepresentation to the public and victims? In the case of the Samarco Dam breach, the FMEA has reportedly been cited in the Brazilian Federal Police inquiries.

There are ways to build tolerance thresholds based on societal consensus and on models. Again, these methodologies do not fall within the scope of the chapter but can be found in specialized books [28]. However, we discuss below an example, originally proposed in 2011, under the form of an attempt to update Whitman & Morgan thresholds.

The 2011 threshold is not necessarily the "true" large-scale societal acceptability, as we used a few examples of 2000–2011 events that (a) caused significant casualties and (b) by the generated reactions, clearly showed the events were not tolerable in G8 countries. Furthermore, by the very nature of the considered events, the 2011 threshold is most likely located near (just higher than) the new upper bound (2011 tolerance/acceptability threshold).

Let us look at examples of accidents and evaluate whether they are considered societally tolerable or not: (1) several dozen traffic-accident casualties per weekend, several times per year, lead the Italian government to invest a large capital in a continuous real-time speed checking and enforcing system (Traffic Tutor) and road safety, as the situation was deemed intolerable; (2) a quake causing 308 casualties (Aquila), thirty years after another catastrophic

http://dx.doi.org/10.5772/66066

one (Irpinia), leads to the conviction of a number of public officers for mass man-slaughter and various other charges (no such reaction for the Irpinia one, thirty years before); (3) a terrorist act (9/11, New York) caused approximately 3000 casualties and the USA "declared war on terrorism;" (4) a quake and a tsunami (Fukushima) with a wave considered to be larger than the MCE caused an evacuation zone of 20 km and then 30 km radius, with very large number of afflicted people (which may become ill in the future), leading Germany and other countries to decide to stop their nuclear energy programs, showing that the event was considered intolerable.

For each one of the examples above, it was possible to define a rate of occurrence (a range of rates); then knowing the range of consequences, it was possible to draft the societal-tolerance thresholds (optimistic, pessimistic) updated to 2000–2011.

It is important to distinguish between location-based risks and societal risks. The first is an expression of the risk exposure for someone who lives or works in a place where a hazardous activity takes place. Societal risk is quite different: it looks at the consequences of mishaps from a very broad point of view of an entire society, possibly physically and emotionally removed from the mishap itself; as such, it is of interest mainly to public administrators. Tolerable risk curves are always project- and owner-specific and indicate the level of risk, which has been deemed acceptable by an owner for a specific project or operation (possibly taking into account public opinion). This means, as an example, that, within large companies, corporate-risk tolerability may differ quite substantially from a branch operation's tolerability.

The development of empirical-estimated tolerability curves requires caution and continuous calibration (see example above); they should always be defined by a group and not by an individual.

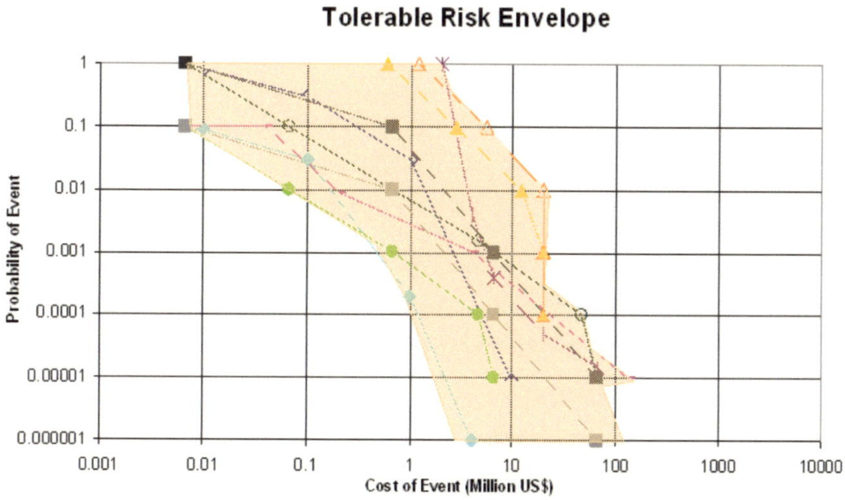

Figure 8. Examples of empirically derived corporate tolerance curves.

In a risk study, great attention must be exerted in ensuring that the acceptability curves are derived for the considered risks: curves derived for hazardous industrial activities cannot be used for natural GHZs (like for typhoons, quakes, or flooding) or business risks. **Figure 8** shows a series of curves derived through discussion with European and North American companies that have been willing to develop their own risk tolerability. In reality, oftentimes, these curves correspond to the perceived tolerability rather than the real "absolute" financial capacity of the company to withstand the occurrence of a damage due to GHZ or other natural or man-made hazards.

4.5.2. Defining manageable vs. unmanageable, strategic vs. tactical & operational risks

Strategic planning is an organization's process of defining its strategy or direction and making decisions on allocating its resources to pursue this strategy. Typically, strategic choices look at three to five years (say 30–50 for GHZs or more), although some extend their vision to 20 years (long term). Because of the time horizon and the nature of the questions dealt, mishaps potentially occurring during the execution of a strategic plan are afflicted by significant uncertainties and may lie very remotely out of the control of management (e.g., of MCEs, quakes, landslide, forest fire, flooding). Those mishaps, in conjunction with their potential consequences, are called "strategic risks."

Tactical planning is short range, emphasizing the current operations of various parts of the organization. Short range is generally defined as a period of time extending about one year or shorter in the future (say ten years for GHZs). Managers use tactical planning to outline what the various parts of the organization must do for the organization to be successful at some point one year or shorter into the future. Tactical plans are usually developed in the areas of production, marketing, personnel, finance, and plant facilities, but have their place with GHZs as well. Because of the time horizon and the nature of the questions dealt, mishaps potentially occurring during the execution of a tactical plan should be covered by moderate uncertainties and may lie closer to the control of management (e.g., of rockfalls and floodings) than strategic ones. Those mishaps, in conjunction with their potential consequences, are called "tactical risks."

Operational planning is the process of linking strategic goals and objectives to tactical goals and objectives. It describes milestones and conditions for success and explains how, or what portion of, a strategic plan will be put into operation during a given operational period (say less than five years in the GHZ space). Operational risks are those arising from the people, systems, and processes through which a company operates and can include those generated by GHZs. A tailings dam failure, an open-pit slide, and a black-out (man-made or GHZ generated) are all operational hazards generating operational risks.

5. Ethical (geoethical) issues in the Anthropocene

Engineering ANPgenic global change is loaded with implicit societal issues to an unprecedented level because of demographic pressure and the raise of public opinion, thanks to the

http://dx.doi.org/10.5772/66066

emergence of the blogosphere. Because of the nonlinear dynamics, perceived or real complex feedbacks, and apparent chaotic dynamics, many claim our ANPcenic systems are difficult to forecast and unintended and counter-intuitive system behavior is likely. In our experience, it is not always so: poor RA tends to mislead people to believe that things may be more complex that what they really are [29, 30].

Academic and popular literatures suggest an agreement that the public's distrust has developed over the past half century as a result of repeated failures to provide adequate and/or accurate risk information to the public. In the public health arena, regulators have traditionally been confronted with the difficult task of allocating risks and benefits; sometimes, they have missed some important risks, and sometimes, they have spent a lot of money and energy on dealing with negligible risks [31]. GHZ will certainly follow this trend as proven by recent "mining/environmental" cases, allowing to measure public skepticism [10, 32]. In fact, "the scientific majority sometimes finds itself pitted against a public opinion which simply does not accept its conclusions" [33].

Meanwhile, over the last five decades or so, the risk management community at large, including engineers and designers performing RAs on their own projects/designs for civil projects, oftentimes in a conflict of interest situations, has settled on representing the results of RAs using misleading methods (see Section 2.3) and has maintained poor communication habits.

The implications of poor risk prioritization for the world industry's balance sheet facing GHZs can be staggering, aside from the possible liabilities. Inaccuracies can lead to mistaken resource allocation and create fuzziness for DMs and the public, thus offer little support to rational decision-making, and lead to public distrust and loss of confidence because of their arbitrariness. It does not come as a surprise then to recognize that, contrary to what is proposed by international codes like ISO 31000, communication and risk approaches are poorly developed through the life of projects and operations (**Figure 9**).

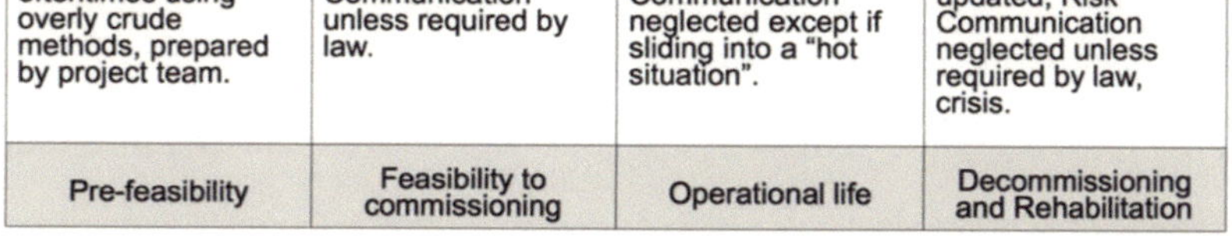

Figure 9. The evolution of RA and communication through a project and operational life. Red-shaded panels indicate poorly performed functions of the RA process.

It is normally accepted that experts disagree in their analyses (e.g., probability or frequency estimates for an event). However, if and when the public disagrees with an expert analysis of risk, they are dismissed as being highly emotional or lacking scientific literacy [34, 35].

This concept is important because while there is an accepted difference in scientific literacy between the public and scientific (GHZs) experts, there is an assumption that the public are ignorant about risks and probabilities and that an increased scientific literacy would help decrease perceived risks [36]. An increase in scientific literacy may in fact increase perceived risks. The question remains as to whether the level of required scientific literacy is "so high that it is difficult to attain and difficult to motivate the public to attain it" [37]. It is simply unrealistic that the average citizen can obtain sufficient scientific (GHZs) literacy to thoroughly tackle any GHZs RAs. Thus, RMs must move their communication approach from that of paternalistically doling out pieces of information supporting their RM approach to partnering with the public [38] to demonstrate that the practices meet socially acceptable levels and practices. Partnering with the public requires effective communication, but more importantly, public consultation and participation. Two vital components of GHZs risk communication are trust and credibility, which corporations and governments must earn [32]. Research must aid risk analysis and policy making by, in part, "improving the communication of risk information among lay people, technical experts, and DMs" [39]. The price to pay if communication and partnering are not improved is never-ending crises, turmoil, boycotts, and possibly revolts.

6. Conclusions

We will close this discussion on Human-Generated Geohazards with an example drawn from a regional-scale study performed a few years ago (**Figure 10**). **Figure 10** displays three selected locations where man-made GHZs were of particular interest: (a) an area where a river had been excessively channelized and increased flooding chances downstream, (b) an industrial area where sludge and waste ponds have breach potential, and (c) a mountainous area where bridges and roads have increased various families of risks. At each site, the "columns" represent the risks due to a certain hazard, split (the three colors) by consequence type (direct, indirect, and social).

The study complied with the requirements described in Section 2.3.2 [10] and was geared toward showing to the public convergent ANPcenic risks deriving from natural, industrial, transportation, and agricultural generated GHZs.

The model satisfied all five roles of system science, namely:

- described the physical world **[Glossary (Section 2.1); System Description (Section 2.2)]**;
- portrayed the results of interactions among a few of its components **[Consequence Analysis (Section 3.2)]**;
- proposed a generic design **[Mitigations (Section 4.5.2 and Literature)]**;

http://dx.doi.org/10.5772/66066

- was a constituent of "science of complexity" as it enlarged the domain of demonstrable results in the service of humanity **[Interdependencies (Section 4.4)]**; and
- was actionable, as it has linguistic clarity and suggested clear direction of actions essential to resolve emergencies **(Decision-Making: Literature).**

The model also addressed some of the Complexity Laws [2], like:

- did not require humans to process more than three components at a time (triadic constraint) **[e.g., Probability (Section 3.1), Consequence, Risk, and Perception (Section 4.)];**
- rendered a parsimonious description of any emergency **[Unified Scale of Consequence (Section 3.2)];**
- addressed the challenge of vertical incoherence as it showed the right aggregated level to decision-makers at different organizational levels **[Interdependencies (Section 4.4), Communication, Information Dashboards (Section 2.3.2), and literature];**
- considered all relevant factors of emergencies in a balanced fashion **(Rational Risk Prioritization: Literature)** [3].

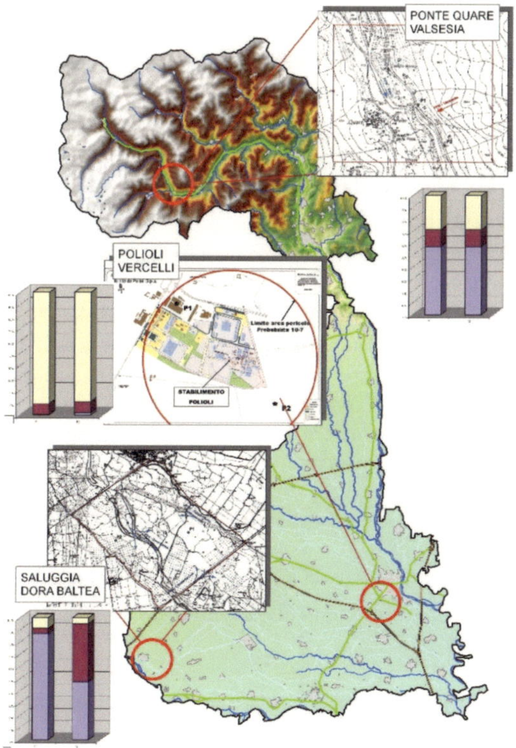

Figure 10. An example of a convergent, ANPcenic-GHZ, quantitative RA at regional scale.

It is time to stop seeking excuses and develop Risk Management 2.0!

Author details

Franco Oboni* and Cesar Oboni

*Address all correspondence to: foboni@riskope.com

Oboni Riskope Associates Inc., Vancouver, B.C., Canada

References

[1] Warfield, J. N., A proposal for systems science. Systems Research and Behavioral Science, 20(6): 507–520, 2003.

[2] Warfield, J. N., Understanding Complexity: Thought and Behavior. AJAR Publishing Company, Palm Harbor, FL, ISBN 0-971-6962-0-9, 2002.

[3] Asproth, V., Håkansson, A., Complexity challenges of critical situations caused by flooding. Emergence: Complexity and Organization, 9(1): 37–43, ISSN 1532-7000, 2007.

[4] Oboni, F., Caldwell, J., Oboni, C., Ten rules for preparing sensible risk assessments. Proceedings of Risk and Resilience Mining Solutions, Infomine, Vancouver, Canada, 2016.

[5] Oboni, F., Oboni, C., Caldwell, J., Risk assessment of the long-term performance of closed tailings, Tailings and Mine Waste 2014, Keystone, Colorado, USA, October 5–8, 2014.

[6] NASA, NASA Systems Engineering Handbook SP-2007-6105. National Aeronautics and Space Administration, Washington, DC, Chapter 6.4, 2007.

[7] Chapman, C., Ward, S., The Probability-Impact Grid – A Tool That Needs Scrapping. How to Manage Project Opportunity and Risk, Chapter 2, pp. 49–51, 3rd Ed., Wiley, West Sussex, United Kingdom, 2011.

[8] Cox, L. A. Jr., What's wrong with risk matrices? Risk Analysis, 28(2), pp. 1–20, 2008.

[9] Hubbard, D., Worse Than Useless. The Most Popular Risk Assessment Method and Why It Doesn't Work. The Failure of Risk Management, Chapter 7, Wiley, West Sussex, United Kingdom, 2009.

[10] Mackenzie Valley Review Boards, Report of Environmental Assessment and Reasons for Decision, Giant Mine Remediation Project, Appendix D, Yellowknife, NWT, Canada, pp. 227–228, June, 2013.

[11] Wilson, A. C., Crouch, E., Risk/Benefit Analysis. Ballinger Publishing Company, Boston, MA, pp. 92–93, 1982.

[12] Taleb, N. N., The Black Swan: The Impact of the Highly Improbable. Random House, 26 ISBN 978-1400063512, New York, United States, 2007.

[13] Government Office for Science, Blackett Review of High Impact Low Probability Risks. Government Office for Science, UK, 2011.

[14] Turner, B., A., Pidgeon, N., F., Man Made Disasters, 2nd ed., Butterworth-Heinemann, Oxford, 1998.

[15] Kahneman, D., Tversky, A., 1979. "Prospect Theory" quoted in Oboni, F. Oboni, C., p. 32 212, Systech, Froideville, Suisse, 2007.

[16] Ang, A. H-S., Tang, W. H., Probability Concepts in Engineering Planning and Design, 34 Vol. I, Wiley, United States, 1975.

[17] Ang, A. H-S., Tang, W. H., Probability Concepts in Engineering Planning and Design, 2 Vol. II, Wiley, United States, 1984.

[18] Ropeik, D., The L'Aquila Verdict: A Judgment Not Against Science, But Against a Failure of Science Communication, Scientific American, October 22, 2012.

[19] Oboni, F., Bourdeau, P. L., Determination of the critical slip surface in stability problems, Proceeding of IVth International Conference on Application of Statistics and Probability in Soil and Structural Engineering, Florence. Università di Firenze (Italy) 1983, Pitagora Editrice, pp 1413–1424, 1983.

[20] Oboni, F., Angelino, C., Moreno, J., Using artificial intelligence in an integrated risk management program for a large alpine landslide, International Conference 'Climate Change – Challenges and Solutions,' Ventnor, Isle of Wight, United Kingdom, May 21–24, 2007.

[21] Appleby, M., Forlin, G., et al., The Law Relating to Emergencies and Disasters. In Tolley's Handbook of Disaster and Emergency Management: Principles and Practice, R. Lakha and T. Moore (Eds.), Butterworth-Heinemann, ISBN 0-40697270-2, Oxford, United Kingdom, 2003.

[22] Federal Office for Civil Protection, KATARISK: disasters and emergencies in Switzerland, A risk assessment from the perspective of the Civil Protection, Bern, Switzerland, August 2003.

[23] Plotnick, L., Gomez, E. A., White., C., Turoff, M., Furthering development of a unified emergency scale using Thurstone's law of comparative judgment: A progress report, Proceedings ISCRAM, Information Systems Department New Jersey Institute of Technology, Newark, NJ, USA, 2007.

[24] Rohn, E., Blackmore, D., A unified localizable emergency events scale. International Journal of Information Systems for Crisis Response Management (IJISCRAM), 1(4), pp. 1–14, 2009.

[25] Klinke, A., Renn, O., Prometheus Unbound, Challenges of Risk Evaluation, Risk Classification and Risk Management, ISBN: 3-932013-95-6, http://dx.doi.org/10.18419/

opus-8558, Center of Technology Assessment in Baden-Württemberg, Germany; 153, 1999.

[26] Whitman, R. V., Evaluating calculated risk in geotechnical engineering. Journal of Geo-engineering Sciences, 110(2), pp. 143–188, 1984.

[27] ANCOLD, International Commission on Large Dams, Australian National Committee. Guidelines on Risk Assessment, Melbourne, Australia, 2003.

[28] Oboni, F., Oboni, C., Improving Sustainability through Reasonable Risk and Crisis Management, ISBN 978-0-9784462-0-8, JSO, Froideville, Switzerland, 2007.

[29] Oboni, F., Oboni, C., Ethics and Transparent Risk Communication Start with Proper Risk Assessment Methodologies, EGU General Assembly 2014, Vienna, May, 2014.

[30] Oboni, F., Oboni, C., Zabolotniuk, S., Can We Stop Misrepresenting Reality to the Public?, CIM 2013, Toronto, 2013.

[31] Bouder, F., Slavin, D., Loefstedt, R. (Eds.), The Tolerability of Risk: A New Framework for Risk Management, Earthscan, New York, United States 2007.

[32] Peters, R. G., Covello, V. T., Mccallum, D. B., The determinants of trust and credibility in environmental risk communication: An empirical study. Risk Analysis, 17(1): 43–54, 1997.

[33] Sjöberg, L., Risk perception by the public and by experts: A dilemma in risk management. Human Ecology Review, 6(2): 1–9, 1999.

[34] Durant, J. R., What is Scientific Literacy? In Science and Culture in Europe, J. R. Durant and J. Gregory (Eds.), pp. 129–137, Science Museum, London, 1993.

[35] Shen, B. S. P., Scientific Literacy and the Public Understanding of Science. In Communication of Scientific Information, S. B. Day (Eds.), pp. 44–52, Karger, Basel, 1975.

[36] Frewer, L., The public and effective risk communication. Toxicology Letters, 149(1): 391–397, 2004.

[37] Frewer, L. J., Howard, C., Hedderley, D., Shepherd, R., What determines trust in information about food related risks? Underlying psychological constructs. Risk Analysis, 16(4): 473–486, 2006.

[38] Fischhoff, B., Risk perception and communication unplugged: Twenty years of process. Risk Analysis, 15(2): 137–145, 1995.

[39] Slovic, P., Risk perception. Science, 236(4799): 280–285, 1987.